The Analysis of Time Series: Theory and Practice

C. CHATFIELD

Lecturer in Statistics,
The University of Bath

CHAPMAN AND HALL

LONDON

First published 1975
by Chapman and Hall Ltd
11 New Fetter Lane, London EC4P 4EE

© 1975 C. Chatfield

Typeset by Preface Ltd, Salisbury, Wiltshire
and printed in Great Britain by
Whitstable Litho, Whitstable, Kent

ISBN 0 412 14180 9

Distributed in the U.S.A.
by Halsted Press, a Division
of John Wiley & Sons, Inc., New York

Library of Congress Catalog Card Number 75-8682

To
Liz

Alice sighed wearily. 'I think you might do something better with the time,' she said, 'than waste it in asking riddles that have no answers.'

'If you knew Time as well as I do,' said the Hatter, 'you wouldn't talk about wasting *it*. It's *him*.'

'I don't know what you mean,' said Alice.

'Of course you don't!' the Hatter said, tossing his head contemptuously. 'I dare say you never even spoke to Time!'

'Perhaps not,' Alice cautiously replied: 'but I know I have to beat time when I learn music.'

'Ah! that accounts for it,' said the Hatter. 'He won't stand beating.'

Alice's Adventures in Wonderland,
Lewis Carroll.

Contents

Preface xii

Abbreviations xiv

1 **Introduction** 1
 1.1 Examples 1
 1.2 Terminology 5
 1.3 Objectives of time-series analysis 6
 1.4 Approaches to time-series analysis 9
 1.5 Review of time-series literature 10

2 **Simple descriptive techniques** 12
 2.1 Types of variation 12
 2.2 Stationary time series 14
 2.3 Time plot 14
 2.4 Transformations 14
 2.5 Analysing series which contain a trend 16
  2.5.1 Curve-fitting 16
  2.5.2 Filtering 17
  2.5.3 Differencing 21
 2.6 Seasonal fluctuations 21
 2.7 Autocorrelation 23
  2.7.1 The correlogram 25
  2.7.2 Interpreting the correlogram 25
 2.8 Other tests of randomness 31
 Exercises 31

3 **Probability models for time series**   33
   3.1 Stochastic processes   33
   3.2 Stationary processes   35
     3.2.1 Second-order stationarity   37
   3.3 The autocorrelation function   37
   3.4 Some useful stochastic processes   39
     3.4.1 A purely random process   39
     3.4.2 Random walk   40
     3.4.3 Moving average processes   41
     3.4.4 Autoregressive processes   44
     3 4.5 Mixed models   50
     3.4.6 Integrated models   51
     3.4.7 The general linear process   52
     3.4.8 Continuous processes   52
   3.5 The Wold decomposition theorem   54
   Exercises   55

4 **Estimation in the time domain**   60
   4.1 Estimating the autocovariance and autocorrelation functions   60
     4.1.1 Interpreting the correlogram   63
     4.1.2 Ergodic theorems   65
   4.2 Fitting an autoregressive process   65
     4.2.1 Estimating the parameters of an autoregressive process   65
     4.2.2 Determining the order of an autoregressive process   69
   4.3 Fitting a moving average process   70
     4.3.1 Estimating the parameters of a moving average process   70
     4.3.2 Determining the order of a moving average process   73
   4.4 Estimating the parameters of a mixed model   73
   4.5 Estimating the parameters of an integrated model   74
   4.6 The Box–Jenkins seasonal model   74

4.7	Residual analysis	76
4.8	General remarks on model building	79
	Exercises	81

5	**Forecasting**	**82**
	5.1 Introduction	82
	5.2 Univariate procedures	84
	5.2.1 Extrapolation of trend curves	84
	5.2.2 Exponential smoothing	85
	5.2.3 Holt–Winters forecasting procedure	87
	5.2.4 Box–Jenkins forecasting procedure	89
	5.2.5 Stepwise autoregression	93
	5.2.6 Other methods	93
	5.3 Multivariate procedures	95
	5.3.1 Multiple regression	95
	5.3.2 Econometric models	97
	5.3.3 Box–Jenkins method	97
	5.4 A comparison of forecasting procedures	98
	5.5 Some examples	101
	5.6 Prediction theory	107
	Exercises	108

6	**Stationary processes in the frequency domain**	**110**
	6.1 Introduction	110
	6.2 The spectral distribution function	111
	6.3 The spectral density function	116
	6.4 The spectrum of a continuous process	119
	6.5 Examples	120
	Exercises	125

7	**Spectral analysis**	**127**
	7.1 Fourier analysis	127
	7.2 A simple sinusoidal model	128
	7.2.1 The Nyquist frequency	131
	7.3 Periodogram analysis	133

	7.3.1 The relationship between the periodogram and the autocovariance function	136
	7.3.2 Properties of the periodogram	137
7.4	Spectral analysis: some consistent estimation procedures	138
	7.4.1 Transforming the truncated autocovariance function	139
	7.4.2 Hanning	142
	7.4.3 Hamming	143
	7.4.4 Smoothing the periodogram	143
	7.4.5 The fast Fourier transform	145
7.5	Confidence intervals for the spectrum	149
7.6	A comparison of different estimation procedures	150
7.7	Analysing a continuous time series	155
7.8	Discussion	158
7.9	An example	165
	Exercises	168
8	**Bivariate processes**	**169**
8.1	Cross-covariance and cross-correlation functions	169
	8.1.1 Examples	171
	8.1.2 Estimation	172
	8.1.3 Interpretation	174
8.2	The cross-spectrum	174
	8.2.1 Examples	178
	8.2.2 Estimation	180
	8.2.3 Interpretation	184
	Exercises	185
9	**Linear systems**	**186**
9.1	Introduction	186
9.2	Linear systems in the time domain	187
	9.2.1 Some examples	188

		9.2.2 The impulse response function	192
		9.2.3 The step response function	192
	9.3	Linear systems in the frequency domain	193
		9.3.1 The frequency response function	193
		9.3.2 Gain and phase diagrams	198
		9.3.3 Some examples	200
		9.3.4 General relation between input and output	203
		9.3.5 Linear systems in series	210
		9.3.6 Design of filters	211
	9.4	Identification of linear systems	213
		9.4.1 Estimating the frequency response function	215
		9.4.2 The Box–Jenkins approach	219
		9.4.3 Systems involving feedback	223
	Exercises		226

10 Some other topics — **228**

Appendix I   The Fourier, Laplace and Z-transforms — 233

Appendix II   The Dirac Delta Function — 238

Appendix III   Covariance — 240

References — 242

Answers to exercises — 252

Author Index — 258

Subject Index — 261

Preface

Time-series analysis is an area of statistics which is of particular interest at the present time. Time series arise in many different areas, ranging from marketing to oceanography, and the analysis of such series raises many problems of both a theoretical and practical nature. I first became interested in the subject as a postgraduate student at Imperial College, when I attended a stimulating course of lectures on time-series given by Dr. (now Professor) G. M. Jenkins. The subject has fascinated me ever since.

Several books have been written on theoretical aspects of time-series analysis. The aim of this book is to provide an introduction to the subject which bridges the gap between theory and practice. The book has also been written to make what is rather a difficult subject as understandable as possible. Enough theory is given to introduce the concepts of time-series analysis and to make the book mathematically interesting. In addition, practical problems are considered so as to help the reader tackle the analysis of real data.

The book assumes a knowledge of basic probability theory and elementary statistical inference (see Appendix III). The book can be used as a text for an undergraduate or postgraduate course in time-series, or it can be used for self tuition by research workers.

Throughout the book, references are usually given to recent readily accessible books and journals rather than to the original attributive references. Wold's (1965) bibliography contains many time series references published before 1959.

PREFACE

In writing the book, I have been struck by the many different definitions and notations used by different writers for different features of time-series. This makes the literature rather confusing. I have adopted what seems to me to be the most sensible definition and notation for each feature, but I have always tried to make it clear how my approach is related to that of other authors.

A difficulty with writing a textbook is that many practical problems contain at least one feature which is 'non-standard' and these cannot all be envisaged in a book of reasonable length. Thus I want to emphasize that the reader who has grasped the basic concepts of time-series analysis, should always be prepared to use his commonsense in tackling a problem. Example 1 of Section 5.5 is a typical situation where common-sense has to be applied and also stresses the fact that the first step in any time-series analysis should always be to plot the data.

I am indebted to many people for helpful comments on earlier drafts, notably Chris Theobald, Mike Pepper, John Marshall, Paul Newbold, David Prothero, Henry Neave, and Professors V. Barnett, D. R. Cox, and M. B. Priestley. Dick Fenton carried out the computing for Section 7.9. I would particularly like to thank Professor K. V. Diprose who made many useful suggestions regarding linear systems and helped me write Section 9.4.3. Of course any errors, omissions or obscurities which remain are entirely my responsibility. The author will be glad to hear from any reader who wishes to make constructive comments.

Finally it is a particular pleasure to thank Mrs. Jean Honebon for typing the manuscript with exceptional efficiency.

School of Mathematics,   Christopher Chatfield,
University of Bath.
December 1974.

Abbreviations

AR	Autoregressive
MA	Moving Average
ARMA	Mixed Autoregressive Moving Average
ARIMA	Autoregressive Integrated Moving Average
ac.f	Autocorrelation function
acv.f.	Autocovariance function
FFT	Fast Fourier Transform
$N(\mu,\sigma^2)$	A normally distributed random variable, mean μ, variance σ^2.
χ^2_ν	A chi-squared random variable with ν degrees of freedom

Notation

∇	The difference operator such that $\nabla x_t = x_t - x_{t-1}$.
B	The backward shift operator such that $Bx_t = x_{t-1}$.
E	Expected value or expectation

CHAPTER 1

Introduction

A time series is a collection of observations made sequentially in time. Examples occur in a variety of fields, ranging from economics to engineering, and methods of analysing time series constitute an important area of statistics.

1.1 Examples

We begin with some examples of the sort of time series which arise in practice

(a) Economic time series　　Many time series arise in economics, including such series as share prices on successive days, export totals in successive months, average incomes in successive months, company profits in successive years; and so on. Fig. 1.1 shows part of the classic Beveridge wheat price index series which consists of the average wheat price in nearly 50 places in various countries measured in successive years from 1500 to 1869. The complete series is given by Anderson (1971). This series is of particular interest to economic historians, and, when analysed (Granger and Hughes, 1971), shows clear evidence of an important cycle with a period of roughly 13.3 years.

(b) Physical time series　　Many types of time series occur in the physical sciences, particularly in meteorology, marine science, and geophysics. Examples are rainfall on successive

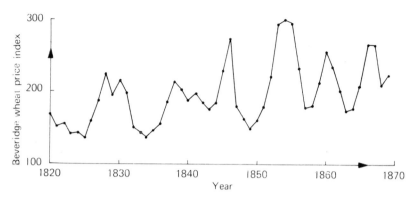

Figure 1.1 *Part of the Beveridge wheat price index series.*

days, and air temperature measured in successive hours, days or months. Fig. 1.2 shows the air temperature at Recife, in Brazil, measured over 10 years, where the individual observations are averaged over periods of one month.

Some mechanical recorders take measurements continuously and produce a continuous trace rather than observations at discrete intervals of time. For example in some laboratories it is important to keep temperature and humidity as constant as possible and so devices are installed to measure these variables continuously. Some examples of continuous traces are given in Chapter 7 in Fig. 7.4.

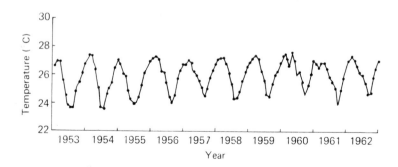

Figure 1.2 *Average air temperature at Recife, Brazil, in successive months.*

2

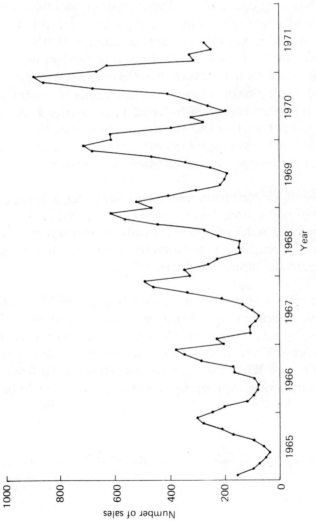

Figure 1.3   Sales of a certain engineering company in successive months.

Continuous recorders often smooth the data to some extent as they are insensitive to high frequency variation.

(c) Marketing time series   The analysis of sales figures in successive weeks or months is an important problem in commerce. Fig. 1.3, taken from Chatfield and Prothero (1973a), shows the sales of an engineering product by a certain company in successive months over a seven-year period. Marketing data have many similarities to economic data. It is often important to forecast future sales so as to plan production. It is also of interest to examine the relationship between sales and other time series such as advertising expenditure in successive time periods.

(d) Demographic time series   Time series occur in the study of population. An example is the population of England and Wales measured annually. Demographers want to predict changes in population for as long as ten or twenty years into the future (e.g. Brass, 1974).

(e) Process control   In process control, the problem is to detect changes in the performance of a manufacturing process by measuring a variable which shows the quality of the process. These measurements can be plotted against time as in Fig. 1.4. When the measurements stray too far from some target value, appropriate corrective action should be

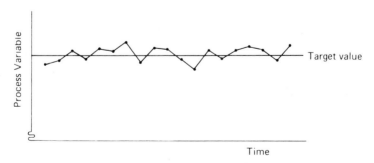

Figure 1.4   A process control chart.

taken to control the process. Special techniques have been developed for this type of time-series problem and the reader is referred to a book on statistical quality control (e.g. Wetherill, 1969).

(f) Binary processes   A special type of time series arises when observations can take one of only two values, usually denoted by 0 and 1 (see Fig. 1.5). Time series of this type, called binary processes, occur particularly in communication theory. For example the position of a switch, either 'on' or 'off', could be recorded as one or zero respectively.

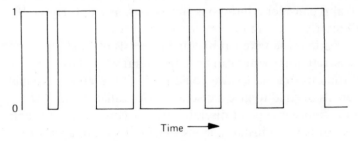

Time ⟶

Figure 1.5   A realization of a binary process.

(g) Point processes   A different type of time series occurs when we consider a series of events occurring 'randomly' in time. For example we could record the dates of major railway disasters. A series of events of this type is often called a *point process* (see Fig. 1.6). For observations of this type, we are interested in the distribution of the number of events occurring in a given time-period and also in the distribution of time intervals between events. Methods of analysing data of this type are discussed by Cox and Lewis (1966), and will not be specifically discussed in this book.

Time ⟶

Figure 1.6   A realization of a point process. (✕ denotes an event.)

1.2 Terminology

A time series is said to be *continuous* when observations are made continuously in time as in Figs. 1.5 and 7.4. The term continuous is used for series of this type even when the measured variable can only take a discrete set of values, as in Fig. 1.5. A time series is said to be *discrete* when observations are taken only at specific times, usually equally spaced. The term discrete is used for series of this type even when the measured variable is a continuous variable.

In this book we are mainly concerned with discrete time series, where the observations are taken at equal intervals, We will also consider continuous time series, and in Chapter 10 will briefly consider discrete time series where the observations are taken at unequal intervals of time.

Discrete time series can arise in several ways. Given a continuous time series, we could read off the values at equal intervals of time to give a discrete series called a *sampled* series. Another type of discrete series occurs when a variable does not have an instantaneous value but we can *aggregate* (or accumulate) the values over equal intervals of time. Examples of this type are exports measured monthly and rainfall measured daily. Finally some time series are inherently discrete, an example being the Financial Times share index at closing time on successive days.

Much statistical theory is concerned with random samples of independent observations. The special feature of time-series analysis is the fact that successive observations are usually *not* independent and that the analysis must take into account the time *order* of the observations. When successive observations are dependent, future values may be predicted from past observations. If a time series can be predicted exactly, it is said to be *deterministic.* But most time series are *stochastic* in that the future is only partly determined by past values. For stochastic series exact predictions are impossible

6

and must be replaced by the idea that future values have a probability distribution which is conditioned by a knowledge of past values.

1.3 Objectives of time-series analysis

There are several possible objectives in analysing a time series. These objectives may be classified as description, explanation, prediction and control, and will be considered in turn.

(a) Description   When presented with a time series, the first step in the analysis is usually to plot the data and to obtain simple descriptive measures of the main properties of the series as described in Chapter 2. For example, looking at Fig. 1.3 it can be seen that there is a regular seasonal effect with sales 'high' in winter and 'low' in summer. It also looks as though annual sales are increasing (i.e. show an upward trend). For some series, the variation is dominated by such 'obvious' features, and a fairly simple model, which only attempts to describe trend and seasonal variation, may be perfectly adequate to describe the variation in the time series. For other series, more sophisticated techniques will be required to provide an adequate analysis. Then a more complex model will be constructed, such as the various types of stochastic process described in Chapter 3.

This book devotes a greater amount of space to the more advanced techniques, but this does not mean that elementary descriptive techniques are unimportant. Anyone who tries to analyse a time series, without plotting it first, is asking for trouble. Not only will a graph show up trend and seasonal variation, but it also enables one to look for 'wild' observations or *outliers* which do not appear to be consistent with the rest of the data. The treatment of outliers is a complex subject in which common-sense is more important

7

than statistical theory, and we will only make a few brief remarks. The 'outlier' may be a perfectly valid observation in which case the model for the time series will need to take it into account. Alternatively, the outlier may be a freak observation arising, for example, when a recording device goes wrong or when a strike severely affects sales. In such cases the outlier needs to be adjusted to its expected value, under normal circumstances, before further analysis of the time series.

Another feature to look for in the graph of the time series is the possible presence of turning points, where, for example, an upward trend has suddenly changed to a downward trend. If there is a turning point, different models may have to be fitted to the two parts of the series.

(b) Explanation   When observations are taken on two or more variables, it may be possible to use the variation in one time series to explain the variation in another series. This may lead to a deeper understanding of the mechanism which generated a given time series.

Multiple regression models may be helpful here. In Chapter 9 we also consider the analysis of what are called *linear systems.* A linear system converts an input series to an output series by a linear operation. Given observations on the input and output to a linear system (see Fig. 1.7), one wants to assess the properties of the linear system. For example it is of interest to see how sea level is affected by temperature and pressure, and to see how sales are affected by price and economic conditions.

(c) Prediction   Given an observed time series, one may want to predict the future values of the series. This is an

Figure 1.7   Schematic representation of a linear system.

8

important task in sales forecasting, and in the analysis of economic and industrial time series. Many writers, including myself, use the terms 'prediction' and 'forecasting' interchangeably, but some authors do not. For example Brown (1963) uses 'prediction' to describe subjective methods and 'forecasting' to describe objective methods, whereas Brass (1974) uses 'forecast' to mean any kind of looking into the future, and 'prediction' to denote a systematic procedure for doing so.

Prediction is closely related to *control* problems in many situations. For example if one can predict that a manufacturing process is going to move off target, then appropriate corrective action can be taken.

(d) Control   When a time series is generated which measures the 'quality' of a manufacturing process, the aim of the analysis may be to control the process. Control procedures are of several different kinds. In statistical quality control, the observations are plotted on control charts and the controller takes action as a result of studying the charts. A more sophisticated control strategy has been described by Box and Jenkins (1970). A stochastic model is fitted to the series, future values of the series are predicted, and then the input process variables are adjusted so as to keep the process on target. Most other contributions to control theory have been made by control engineers (see Chapter 10). Then the 'controller' may be an analog or digital computer.

1.4 Approaches to time-series analysis

This book will describe various approaches to time-series analysis. In Chapter 2, we will describe simple descriptive techniques, which consist of plotting the data, looking for trends, seasonal fluctuations, and so on. Chapter 3 introduces a variety of probability models for time series, while Chapter 4 discusses ways of fitting these models to time series. The major diagnostic tool which is used in Chapter 4 is a function called the *autocorrelation* function which helps to describe the evolution of a process through time. Inference based on this function is often called an analysis in the *time domain*.

Chapter 5 discusses a variety of forecasting procedures. This chapter is not a prerequisite for the rest of the book and the reader may, if he wishes, proceed from Chapter 4 to Chapter 6.

Chapter 6 introduces a function called the *spectral density* function which describes how the variation in a time series may be accounted for by cyclic components at different frequencies. Chapter 7 shows how to estimate this function, a procedure which is called *spectral analysis*. Inference based on the spectral density function is often called an analysis in the *frequency domain*.

Chapter 8 discusses the analysis of two time series, while Chapter 9 extends this work by considering linear systems in which one series is regarded as the input, while the other series is regarded as the output.

Finally Chapter 10 gives a brief introduction to some more specialized topics.

1.5 Review of time-series literature

This section gives a brief review of the literature on time series. Most books on time series, such as Grenander and Rosenblatt (1957), Quenouille (1957), Hannan (1960 and 1970), Anderson (1971) and Brillinger (1975), are mainly concerned with mathematical theory. Of these, the books by Hannan (1970) and Anderson (1971) are particularly valuable reference works for the more advanced student.

The book by Blackman and Tukey (1959) was for several years the principal source of practical information, even though it was written primarily for communications engineers. Two more recent books which the reader will find useful are Jenkins and Watts (1968), which provides a thorough treatment of spectral analysis, and Granger and Hatanaka (1964), which is concerned with the analysis of economic time series. Other books which the reader may find useful are Volume 3 of Kendall and Stuart (1966), Kendall (1973), and Bendat and Piersol (1971), the latter being written primarily for engineers.

The book by Box and Jenkins (1970) is an important contribution to the time-series literature. It describes the approach to time-series analysis, forecasting and control, which has been developed by these authors since about 1960 and which is based on a particular class of linear stochastic models. An introduction to their approach is given in Chapters 3, 4 and 5.

Some other useful books worth noting include Wold (1965), which gives a bibliography on time series and stochastic processes up to 1959, and Rosenblatt (1963) and Harris (1967) which are collections of papers on time series. Robinson's (1967) book on multichannel time-series analysis is a contribution to computer science rather than statistics. The book by Otnes and Enochson (1972) is written from an engineering point of view and gives some useful material on spectral analysis, but the reader should beware of misprints. Whittle's (1963) book is a mathematical treatise on control theory. This subject, which is closely related to time series, is briefly introduced in Chapter 10, but is otherwise outside the scope of this book. Note that Wiener's (1949) classic work on control theory, which is of considerable historical interest, has recently been re-issued with the rather misleading title of 'Time Series'.

CHAPTER 2

Simple Descriptive Techniques

Statistical techniques for analysing time series, range from the straightforward to the very sophisticated. This chapter describes some fairly simple techniques which will often detect the main properties of a given series. More sophisticated techniques, such as spectral analysis, must await a study of the theory of stationary time series.

2.1 Types of variation

Traditional methods of time-series analysis are mainly concerned with decomposing a series into a trend, a seasonal variation, and other 'irregular' fluctuations. This approach is not always the best approach (Kendall and Stuart, 1966, p. 349) but is still often useful.

These different sources of variation will now be described in more detail.

(a) Seasonal effect Many time series, such as sales figures and temperature readings, exhibit a variation which is annual in period. This type of variation is easy to understand, and we shall see that it can be measured or removed from the data to give de-seasonalized data.

(b) Other cyclic changes Apart from seasonal effects, some time series exhibit variation at a fixed period due to

some other physical cause. An example is daily variation in temperature. In addition some time series exhibit oscillations which do not have a fixed period but which are predictable to some extent. For example economic data may be affected by business cycles with a period varying between about 5 and 7 years. Granger and Hughes (1971) found a cycle of period 13 years in the Beveridge annual time-series index of wheat prices in Western and Central Europe mentioned in Section 1.1(a).

(c) Trend   This may be loosely defined as 'long term change in the mean'. A difficulty with this definition is deciding what is meant by 'long term'. For example, climatic variables sometimes exhibit cyclic variation over a very long time-period such as 50 years. If one just had 20 years data, this long term oscillation would appear to be a trend, but if several hundred years data were available, the long-term oscillation would be visible. Nevertheless in the short term it may still be more meaningful to think of such a long-term oscillation as a trend. Thus in speaking of a 'trend', we must take into account the number of observations available and make a subjective assessment of what is 'long term'. Granger (1966) defines 'trend in mean' as comprising all frequency components whose wavelength exceeds the length of the observed time series.

(d) Other irregular fluctuations   After trend and cyclic variations have been removed from a set of data, we are left with a series of residuals, which may or may not be 'random'. We shall examine various techniques for analysing series of this type to see if some of the apparently irregular variation may be explained in terms of probability models, such as *moving average* or *autoregressive* models which will be introduced in Chapter 3. Alternatively we can see if any cyclic variation is still left in the residuals.

2.2 Stationary time-series

A mathematical definition of a *stationary* time series will be given later on. However it is now convenient to introduce the idea of stationarity from an intuitive point of view. Broadly speaking a time series is said to be stationary if there is no systematic change in mean (no trend), if there is no systematic change in variance, and if strictly periodic variations have been removed.

Most of the probability theory of time series is concerned with stationary time series, and for this reason time-series analysis often requires one to turn a non-stationary series into a stationary one so as to use this theory. For example one may remove the trend and seasonal variation from a set of data and then try to model the variation in the residuals by means of a stationary stochastic process.

2.3 Time plot

The first step in analysing a set of data is to plot the observations against time. This will often show up the most important properties of a series. Features such as trend, seasonality, discontinuities, will usually be visible if present in the series.

If the series is approximately stationary, it will be useful to compute the mean and standard deviation of the observations.

2.4 Transformations

Plotting the data may indicate if it is desirable to transform the values of the observed variable. The two main reasons for making a transformation are

(a) To stabilize the variance   If there is trend in the series and the variance appears to increase with the mean then it

14

may be advisable to transform the data. In particular if the standard deviation is directly proportional to the mean, a logarithmic transformation is appropriate.

(b) To make the seasonal effect additive   If there is a trend in the series and the size of the seasonal effect appears to increase with the mean then it may be advisable to transform the data so as to make the seasonal effect constant. In particular if the size of the seasonal effect is directly proportional to the mean, then the seasonal effect is said to be multiplicative and a logarithmic transformation is appropriate to make the effect additive. However this transformation will only stabilize the variance if the error term is *also* thought to be multiplicative, a point which is sometimes overlooked.

Three seasonal models in common use are:

A) $x_t = m_t + s_t + \epsilon_t$

B) $x_t = m_t s_t + \epsilon_t$

C) $x_t = m_t s_t \epsilon_t$

where x_t = observation at time t

m_t = current mean

s_t = seasonal effect

ϵ_t = random error.

Model A is completely additive and no transformation is required. Model C is completely multiplicative and a logarithmic transformation is certainly appropriate. Model B has multiplicative seasonality and additive error and the relative size of these effects determines whether a transformation is desirable. Further difficulties arise in practice if the seasonal and/or error terms are neither multiplicative nor additive. For example the seasonal effect may increase with the mean but not at such a fast rate so that it is 'midway between multiplicative and additive'. Some *ad hoc* method may be used to deal with this situation, or alternatively, one of the family of transformations proposed by Box and Cox (1964) may be appropriate.

2.5 Analysing series which contain a trend

The analysis of a time series which exhibits 'long term' change in mean depends on whether one wants to (a) measure the trend and/or (b) remove the trend in order to analyse local fluctuations. With seasonal data, it is a good idea to start by calculating successive yearly averages as these will provide a simple description of the underlying trend. An approach of this type is sometimes perfectly adequate, particularly if the trend is fairly small, but sometimes a more sophisticated approach is desired and then the following techniques can be considered.

2.5.1 Curve-fitting
A traditional method of dealing with non-seasonal data which contains a trend, particularly yearly data is to fit a simple function such as a polynomial curve (linear, quadratic, cubic, etc.) or a Gompertz curve (see Gregg *et al.*, 1964, Harrison and Pearce, 1972). The Gompertz curve is given by

$$\log x_t = a - br^t$$

where a, b, r are parameters with $0 < r < 1$. Another curve which can be used for S-shaped data, where the values are expected to approach an asymptotic value as $t \to \infty$, is the logistic curve given by

$$x_t = a/[1 + b \exp(-ct)]$$

where a,b,c are parameters and $x_t \to a$ as $t \to \infty$. Button (1974) discusses the use of this curve for forecasting car ownership.

For all curves of this type, the fitted function provides a measure of the trend, and the residuals provide an estimate of local fluctuations, where the residuals are the differences between the observations and the corresponding values of the fitted curve.

2.5.2 Filtering A second procedure for dealing with a trend is to use a *linear filter* which converts one time series, $\{x_t\}$, into another, $\{y_t\}$, by the linear operation

$$y_t = \sum_{r=-q}^{+s} a_r x_{t+r}$$

where $\{a_r\}$ are a set of weights. In order to smooth out local fluctuations and estimate the local mean, we should clearly choose the weights so that $\Sigma a_r = 1$, and then the operation is often referred to as a *moving average*. Moving averages are discussed in detail by Kendall (1973, Chapters 3 and 4), and we will only provide a brief introduction. Moving averages are often symmetric with $s = q$ and $a_j = a_{-j}$. The simplest example of a symmetric smoothing filter is the simple moving average for which $a_r = 1/(2q + 1)$ for $r = -q, \ldots, +q$, and $Sm(x_t) =$ smoothed value of x_t

$$= \frac{1}{2q + 1} \sum_{i=t-q}^{t+q} x_i.$$

The simple moving average is not generally recommended by itself for measuring trend, although it can be useful for removing seasonal variation. Another example is provided by the case where the $\{a_r\}$ are successive terms in the expansion of $(\frac{1}{2} + \frac{1}{2})^{2q}$. Thus when $q = 1$, the weights are $a_{-1} = a_{+1} = \frac{1}{4}$, $a_0 = \frac{1}{2}$. As q gets large, the weights approximate to a normal curve.

A third example is Spencer's 15 point moving average, which is used for smoothing mortality statistics to get life-tables (Tetley, 1946). This covers 15 consecutive points with $q = 7$ and the weights are

$$\frac{1}{320} [-3, -6, -5, 3, 21, 46, 67, 74, \ldots].$$

As the filter is symmetric, $a_1, \ldots, a_7$ are equal to $a_{-1}, \ldots, a_{-7}$.

Another general approach is to fit a polynomial curve, not to the whole series, but to fit polynomials of the same order to different parts. For example one could fit a cubic polynomial to the first $(2q + 1)$ data points, where $q \geqslant 2$, use this polynomial to determine the interpolated value at the middle of the range (i.e. $t = q + 1$), then repeat the operation using the $(2q + 1)$ data points from $t = 2$ to $t = (2q + 2)$ and so on. An alternative class of piece-wise polynomials, is the class of *spline* functions (see Wold, 1974).

Suppose we have N observations, $x_1, \ldots, x_N$. For a symmetric filter, we can calculate $\text{Sm}(x_t)$ for $t = q + 1$ up to $t = N - q$, but as a consequence lose $2q$ pieces of data. In some situations this is not important but in other situations it is particularly important to get smoothed values up to $t = N$. Kendall and Stuart (1966) discuss the end-effects problem. One can either project the smoothed values by eye, fit a polynomial to the last few smoothed values, or alternatively use an asymmetric filter such as

$$\text{Sm}(x_t) = \sum_{j=0}^{q} a_j x_{t-j}.$$

The popular technique known as *exponential smoothing* (see also Section 5.2.2) takes $q = \infty$, and

$$a_j = \alpha(1 - \alpha)^j$$

where α is a constant such that $0 < \alpha < 1$.

Having estimated the trend, we can look at the local fluctuations by examining

$$\text{Res}(x_t) = \text{residual from smoothed value}$$
$$= x_t - \text{Sm}(x_t)$$
$$= \sum_{r=-q}^{+s} b_r x_{t+r}.$$

This is also a linear filter, and if

$$\text{Sm}(x_t) = \sum_{r=-q}^{+s} a_r x_{t+r}.$$

with $\Sigma a_r = 1$, then $\Sigma b_r = 0$, $b_0 = 1 - a_0$, and $b_r = -a_r$ for $r \neq 0$.

How do we choose the appropriate filter? The answer to this question really requires considerable experience plus a knowledge of the frequency aspects of time-series analysis which will be discussed in later chapters. As the name implies, filters are usually designed to produce an output with emphasis on variation at particular frequencies. For example to get smoothed values we want to remove the local fluctuations which constitute what is called the high-frequency variation. In other words we want what is called a *low-pass* filter. To get $\text{Res}(x_t)$, we want to remove the long-term fluctuations or the low-frequency variation. In other words we want what is called a *high-pass* filter. The Slutsky (or Slutsky-Yule) effect is related to this problem. Slutsky showed that by operating on a completely random series with both averaging and differencing procedures one could induce sinusoidal variation in the data, and he went on to suggest that apparently periodic behaviour in some economic time series might be accounted for by the smoothing procedures used to form the data. We will return to this question later.

FILTERS IN SERIES   Very often a smoothing procedure is carried out in two or more stages – so that one has in effect several linear filters in series. For example two filters in series may be represented as follows:

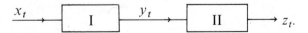

Filter I with weights $\{a_{j1}\}$ acts on $\{x_t\}$ to produce $\{y_t\}$.
Filter II with weights $\{a_{j2}\}$ acts on $\{y_t\}$ to produce $\{z_t\}$.

Now $z_t = \sum\limits_{j} a_{j2}y_{t+j}$

$\qquad = \sum\limits_{j} a_{j2} \sum\limits_{r} a_{r1} x_{t+j+r}$

$\qquad = \sum c_j x_{t+j}$

where

$$c_j = \sum\limits_{r} a_{r1} a_{(j-r)2}$$

are the weights for the overall filter. The weights $\{c_j\}$ are obtained by a procedure called *convolution* and we write

$$\{c_j\} = \{a_{r1}\} \;\ast\; \{a_{r2}\}$$

where * represents convolution.

For example the filter $(\frac{1}{4}, \frac{1}{2}, \frac{1}{4})$ may be written as

$$(\tfrac{1}{4}, \tfrac{1}{2}, \tfrac{1}{4}) = (\tfrac{1}{2}, \tfrac{1}{2}) \ast (\tfrac{1}{2}, \tfrac{1}{2}).$$

Given a series $x_1, \ldots, x_N$, this smoothing procedure is best done in three stages by adding successive pairs of observations twice and then dividing by 4, as follows:

Observations	Stage I	Stage II	Stage III
x_1			
	$x_1 + x_2$		
x_2		$x_2 + 2x_2 + x_3$	$(x_1 + 2x_2 + x_3)/4$
	$x_2 + x_3$		.
x_3		$x_2 + 2x_3 + x_4$	.
	$x_3 + x_4$		.
x_4		$x_3 + 2x_4 + x_5$	.
	$x_4 + x_5$	.	.
x_5		.	.
	$x_5 + x_6$	.	.
x_6	.	.	.
.	.	.	.

The Spencer 15 point moving average is actually a convolution of 4 filters, namely

$$(\tfrac{1}{4}, \tfrac{1}{4}, \tfrac{1}{4}, \tfrac{1}{4}) \ast (\tfrac{1}{4}, \tfrac{1}{4}, \tfrac{1}{4}, \tfrac{1}{4}) \ast (\tfrac{1}{5}, \tfrac{1}{5}, \tfrac{1}{5}, \tfrac{1}{5}, \tfrac{1}{5}) \ast (-\tfrac{3}{4}, \tfrac{3}{4}, 1, \tfrac{3}{4}, -\tfrac{3}{4}).$$

2.5.3 Differencing   A special type of filtering, which is particularly useful for removing a trend, is simply to difference a given time series until it becomes stationary. This method is particularly stressed by Box and Jenkins (1970). For non-seasonal data, first-order differencing is usually sufficient to attain apparent stationarity, so that the new series $\{y_1, \ldots, y_{N-1}\}$ is formed from the original series $\{x_1, \ldots, x_N\}$ by

$$y_t = x_{t+1} - x_t = \nabla x_{t+1}$$

First-order differencing is now widely used in economics. Occasionally second-order differencing is required using the operator ∇^2 where

$$\nabla^2 x_{t+2} = \nabla x_{t+2} - \nabla x_{t+1} = x_{t+2} - 2x_{t+1} + x_t.$$

2.6 Seasonal fluctuations

The analysis of time series which exhibit seasonal fluctuations depends on whether one wants to

(a) measure the seasonals

and/or

(b) eliminate them.

For series showing little trend, it is usually adequate to simply calculate the average for each month (or quarter, or 4-week period etc.) and compare it with the overall average figure, either as a difference or as a ratio.

For series which do contain a substantial trend, a more sophisticated approach may be required. With monthly data, the commonest way of eliminating the seasonal effect is to calculate

$$\text{Sm}(x_t) = \frac{\frac{1}{2}x_{t-6} + x_{t-5} + x_{t-4} + \ldots + x_{t+5} + \frac{1}{2}x_{t+6}}{12}.$$

Note that the sum of the coefficients is 1. A simple moving average cannot be used as this would span 12 months and would not be centred on an integer value of t. A simple moving average over 13 months cannot be used, as this would give twice as much weight to the month appearing at both ends. Durbin (1963) discusses this filter and simple ways of doing the computation. For quarterly data, the seasonal effect can be eliminated by calculating

$$\mathrm{Sm}(x_t) = \frac{\frac{1}{2}x_{t-2} + x_{t-1} + x_t + x_{t+1} + \frac{1}{2}x_{t+2}}{4}.$$

For 4-weekly data, one *can* use a simple moving average over 13 successive observations.

The seasonal effect can be estimated by $x_t - \mathrm{Sm}(x_t)$ or $x_t/\mathrm{Sm}(x_t)$ depending on whether the seasonal effect is thought to be additive or multiplicative. If the seasonal variation stays roughly the same size regardless of the mean level, then it is said to be additive, but if it increases in size in direct proportion to the mean level, then it is said to be multiplicative. The time plot should be examined to see which description is the better.

A check should be made that the seasonals are reasonably stable, and then the average monthly (or quarterly, etc.) effects can be calculated. If necessary a transformation should be made – as previously discussed.

A seasonal effect can also be eliminated by differencing (see Sections 4.6, 5.2.4; Box and Jenkins, 1970). For example with monthly data one can employ the operator ∇_{12} where

$$\nabla_{12} \, x_t = x_t - x_{t-12}.$$

A more complicated moving average procedure is described by Shiskin *et al.* (1967).

2.7 Autocorrelation

An important guide to the properties of a time series is provided by a series of quantities called sample autocorrelation coefficients, which measure the correlation between observations at different distances apart. These coefficients often provide insight into the probability model which generated the data. We assume that the reader is familiar with the ordinary correlation coefficient, namely that given N pairs of observations on two variables x and y, the correlation coefficient is given by

$$r = \frac{\Sigma(x_i - \bar{x})(y_i - \bar{y})}{\sqrt{[\Sigma(x_i - \bar{x})^2 \Sigma(y_i - \bar{y})^2]}} \tag{2.1}$$

A similar idea can applied to time series to see if successive observations are correlated.

Given N observations $x_1, \ldots, x_N$, on a discrete time series we can form $(N-1)$ pairs of observations, namely (x_1, x_2), $(x_2, x_3), \ldots, (x_{N-1}, x_N)$. Regarding the first observation in each pair as one variable, and the second observation as a second variable, the correlation coefficient between x_t and x_{t+1} is given by

$$r_1 = \frac{\sum_{t=1}^{N-1}(x_t - \bar{x}_{(1)})(x_{t+1} - \bar{x}_{(2)})}{\sqrt{[\sum_{t=1}^{N-1}(x_t - \bar{x}_{(1)})^2 \sum_{t=1}^{N-1}(x_{t+1} - \bar{x}_{(2)})^2]}} \tag{2.2}$$

by analogy with equation (2.1), where

$$\bar{x}_{(1)} = \sum_{t=1}^{N-1} x_t/(N-1)$$

is the mean of the first $(N-1)$ observations and

$$\bar{x}_{(2)} = \sum_{t=2}^{N} x_t/(N-1)$$

is the mean of the last $(N-1)$ observations.

As the coefficient given by equation (2.2) measures correlation between successive observations it is called an *autocorrelation* coefficient or serial correlation coefficient.

For N reasonably large, r_1 is approximately given by

$$r_1 = \frac{\displaystyle\sum_{t=1}^{N-1} (x_t - \bar{x})(x_{t+1} - \bar{x})/(N-1)}{\displaystyle\sum_{t=1}^{N} (x_t - \bar{x})^2 / N} \tag{2.3}$$

where $\bar{x} = \displaystyle\sum_{t=1}^{N} x_t / N$

$\qquad$ = overall mean.

Some authors also drop the factor $N/(N-1)$ which is close to one for large N to give

$$r_1 = \frac{\displaystyle\sum_{t=1}^{N-1} (x_t - \bar{x})(x_{t+1} - \bar{x})}{\displaystyle\sum_{t=1}^{N} (x_t - \bar{x})^2} \tag{2.4}$$

and this is the form that will be used in this book.

In a similar way we can find the correlation between observations a distance k apart, which is given by

$$r_k = \frac{\displaystyle\sum_{t=1}^{N-k} (x_t - \bar{x})(x_{t+k} - \bar{x})}{\displaystyle\sum_{t=1}^{N} (x_t - \bar{x})^2} \tag{2.5}$$

This is called the autocorrelation coefficient at lag k.

In practice the autocorrelation coefficients are usually calculated by computing the series of autocovariance coefficients, $\{c_k\}$, which we define by analogy with the usual covariance formula as

$$c_k = \frac{1}{N} \sum_{t=1}^{N-k} (x_t - \bar{x})(x_{t+k} - \bar{x}) \tag{2.6}$$

$$= \text{autocovariance coefficient at lag } k.$$

We then compute

$$r_k = c_k/c_0 \tag{2.7}$$

for $k = 1, 2, \ldots, m$, where $m \leqslant N$. There is usually little point in calculating r_k for values of k greater than about $N/4$.

Note that some authors suggest

$$c_k = \frac{1}{N-k} \sum_{t=1}^{N-k} (x_t - \bar{x})(x_{t+k} - \bar{x})$$

rather than equation 2.6, but there is little difference for large N.

2.7.1 The correlogram   A useful aid in interpreting a set of autocorrelation coefficients is a graph called a correlogram in which r_k is plotted against the lag k. Examples are given in Figs. 2.1–2.3. A visual inspection of the correlogram is often very helpful.

2.7.2 Interpreting the correlogram   Interpreting the meaning of a set of autocorrelation coefficients is not always easy. Here we offer some general advice.

(a) A random series   If a time series is completely random, then for large $N, r_k \cong 0$ for all non-zero values of k. In fact we will see later that for a random time series r_k is approximately $N(0, 1/N)$, so that, if a time series is random, 19 out of 20 of the values of r_k can be expected to lie between $\pm 2/\sqrt{N}$. However if one plots say the first 20 values of r_k then one can expect to find one 'significant' value on average even when the time series really is random. This

spotlights one of the difficulties in interpreting the correlogram, in that a large number of coefficients is quite likely to contain one (or more) 'unusual' results, even when no real effects are present. (See also Section 4.1.)

(b) Short-term correlation   Stationary series often exhibit short-term correlation characterized by a fairly large value of r_1 followed by 2 or 3 more coefficients which, while significantly greater than zero, tend to get successively smaller. Values of r_k for longer lags tend to be approximately zero. An example of such a correlogram is shown in Fig. 2.1. A time series which gives rise to such a correlogram, is one

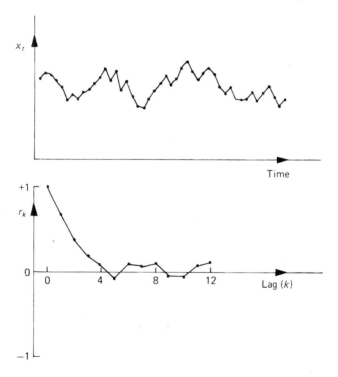

Figure 2.1   A time series showing short-term correlation together with its correlogram

for which an observation above the mean tends to be followed by one or more further observations above the mean, and similarly for observations below the mean. A model called an autoregressive model, may be appropriate for series of this type.

(c) Alternating series   If a time series has a tendency to alternate, with successive observations on different sides of the overall mean, then the correlogram also tends to alternate. The value of r_1 will be negative. However the value of r_2 will be positive as observations at lag 2 will tend to be on the same side of the mean. A typical alternating time series together with its correlogram is shown in Fig. 2.2.

(d) Non-stationary series   If a time series contains a trend, then the values of r_k will not come down to zero except for very large values of the lag. This is because an observation on

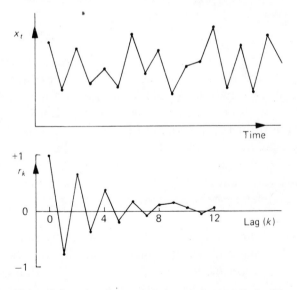

Figure 2.2   An alternating time series together with its correlogram.

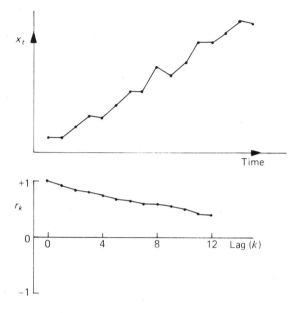

*Figure 2.3   A non-stationary time series together
with its correlogram.*

one side of the overall mean tends to be followed by a large
number of further observations on the same side of the mean
because of the trend. A typical non-stationary time series
together with its correlogram is shown in Fig. 2.3. Little can
be inferred from a correlogram of this type as the trend
dominates all other features. In fact the sample
autocorrelation function, $\{r_k\}$, should only be calculated for
stationary time series (see Chapters 3 and 4) and so any trend
should be removed before calculating $\{r_k\}$.

(e) Seasonal fluctuations   If a time series contains a
seasonal fluctuation, then the correlogram will also exhibit an
oscillation at the same frequency. For example with monthly
observations, r_6 will be 'large' and negative, while r_{12} will be
'large' and positive. In particular if x_t follows a sinusoidal

28

pattern, then so does r_k. For example, if

$$x_t = a \cos t\omega$$

where a is a constant and the frequency ω is such that $0 < \omega < \pi$, then it can easily be shown that (see exercise 3)

$$r_k \cong \cos k\omega \text{ for large } N.$$

Fig. 2.4(a) shows the correlogram of the monthly air temperature data shown in Fig. 1.2

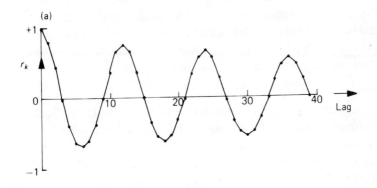

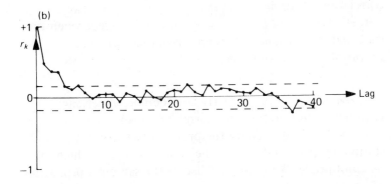

Figure 2.4   The correlogram of monthly observations on air temperature at Recife (a) for the raw data, (b) for the seasonally adjusted data.

The sinusoidal pattern of the correlogram is clearly evident, but for seasonal data of this type the correlogram provides little extra information as the seasonal pattern is clearly evident in the time plot of the data.

If the seasonal variation is removed from seasonal data, then the correlogram may provide useful information. The seasonal variation was removed from the air temperature data by the simple procedure of calculating the 12 monthly averages and subtracting the appropriate one from each individual observation. The correlogram of the resulting series (Fig. 2.4(b)) shows that the first three coefficients are significantly different from zero. This indicates short-term correlation in that a month which is colder than the average for that month will tend to be followed by one or two further months which are colder than average.

(f) Outliers   If a time series contains one or more outliers, the correlogram may be seriously affected and it is usually advisable to adjust outliers as described in Section 1.3. For example, if there is one outlier in the time series and it is not adjusted, then the plot of x_t against x_{t+k} will contain *two* 'extreme' points which will tend to depress the sample correlation coefficients towards zero. If there are two outliers, this effect is even more noticeable except when the lag equals the distance between the outliers when a spuriously large correlation may occur.

(g) General remarks   Clearly considerable experience is required in interpreting autocorrelation coefficients. In addition we need to study the probability theory of stationary series and discuss the classes of model which may be appropriate. We must also discuss the sampling properties of r_k. These topics will be covered in the next two chapters and we shall then be in a better position to interpret the correlogram of a given time series.

2.8 Other tests of randomness

In most cases, a visual examination of the graph of a time series is enough to see that it is *not* random. However it is occasionally desirable to test a stationary time series for 'randomness'. In other words one wants to test if $x_1, \ldots, x_N$ could have arisen in that order by chance by taking a simple random sample size N from a population with unknown characteristics. A variety of tests exist for this purpose and they are described by Kendall and Stuart (1966, Section 45.15) and by Kendall (1973; Chapter 2). For example one can examine the number of times there is a local maximum or minimum in the time series. A local maximum is defined to be any observation x_t such that $x_t > x_{t-1}$ and $x_t > x_{t+1}$. A converse definition applies to local minima. If the series really is random one can work out the expected number of turning points and compare it with the observed value. Tests of this type will not be described here. I have always found it convenient to simply examine the correlogram (and possibly the spectral density function) of a given time series to see if it is random.

Having fitted a model to a non-random series, one often wants to see if the residuals are random. Testing residuals for randomness is a somewhat different problem and will be discussed in Section 4.7.

Exercises

1. The following data show the sales of company X in successive 4-week periods over 1967–1970.

	I	II	III	IV	V	VI	VII	VIII	IX	X	XI	XII	XIII
1967	153	189	221	215	302	223	201	173	121	106	86	87	108
1968	133	177	241	228	283	255	238	164	128	108	87	74	95
1969	145	200	187	201	292	220	233	172	119	81	65	76	74
1970	111	170	243	178	248	202	163	139	120	96	95	53	94

   (a) Plot the data.

   (b) Assess the trend and seasonal effects.

2. Sixteen successive observations on a stationary time series are as follows:-

1.6, 0.8, 1.2, 0.5, 0.9, 1.1, 1.1, 0.6, 1.5, 0.8, 0.9, 1.2, 0.5, 1.3, 0.8, 1.2

   (a) Plot the observations.

   (b) Looking at the graph, guess an approximate value for the autocorrelation coefficient at lag 1.

   (c) Plot x_t against x_{t+1}, and again try to guess the value of r_1.

   (d) Calculate r_1.

3.   If $x_t = a \cos t\omega$ where a is a constant and ω is a constant in $(0, \pi)$, show that $r_k \to \cos k\omega$ as $N \to \infty$.

4.   The first ten sample autocorrelation coefficients of 400 'random' numbers are $r_1 = 0.02$, $r_2 = 0.05$, $r_3 = -0.09$, $r_4 = 0.08$, $r_5 = -0.02$, $r_6 = 0.00$, $r_7 = 0.12$, $r_8 = 0.06$, $r_9 = 0.02$, $r_{10} = -0.08$. Is there any evidence of non-randomness?

CHAPTER 3

Probability Models for Time Series

3.1 Stochastic processes

This chapter describes a variety of probability models for time series, which are collectively called stochastic processes. Most physical processes in the real world involve a random or stochastic element in their structure, and a stochastic process can be described as 'a statistical phenomenon that evolves in time according to probabilistic laws'. Well-known examples are the length of a queue, the size of a bacterial colony, and the air temperature on successive days at a particular site. Many authors use the term 'stochastic process' to describe both the real physical process and a mathematical model of it. The word 'stochastic', which is of Greek origin, is used to mean 'pertaining to chance', and many writers use 'random process' as a synonym for stochastic process.

Mathematically, a stochastic process may be a defined as a collection of random variables $\{X(t), t \in T\}$, where T denotes the set of time-points at which the process is defined. We will denote the random variable at time t by $X(t)$ if T is continuous (usually $-\infty < t < \infty$), and by X_t if T is discrete (usually $t = 0, \pm 1, \pm 2, \ldots$). Thus a stochastic process is a collection of random variables which are ordered in time. For a single outcome of the process we have only one observation on each random variable and these values evolve in time according to probabilistic laws.

THE ANALYSIS OF TIME SERIES

The theory of stochastic processes has been extensively developed. In this chapter we concentrate on those aspects which are particularly relevant for time-series analysis. Further relevant aspects of stochastic processes may be found in many books including Cox and Miller (1968, especially Chapter 7), Bartlett (1966), Parzen (1962), Yaglom (1962), which is mathematically more advanced, and Papoulis (1965) which is written primarily for engineers.

In most statistical problems we are concerned with estimating the properties of a population from a sample. In time-series analysis, however, it is often impossible to make more than one observation at a given time so that we only have one observation on the random variable at time t. Nevertheless we may regard the observed time series as just one example of the infinite set of time series which might have been observed. This infinite set of time series is sometimes called the *ensemble*. Every member of the ensemble is a possible *realization* of the stochastic process. The observed time series can be thought of as one particular realization, and will be denoted by $x(t)$ for $(0 \leqslant t \leqslant T)$ if observations are continuous, and by x_t for $t = 1, \ldots, N$, if observations are discrete.

Because there is only a notional population, time-series analysis is essentially concerned with evaluating the properties of the probability model which generated the observed time series.

One way of describing a stochastic process is to specify the joint probability distribution of $X_{t_1}, \ldots, X_{t_n}$ for any set of times $t_1, \ldots, t_n$ and any value of n. But this is rather complicated and is not usually attempted in practice. A simpler, more useful way of describing a stochastic process is to give the *moments* of the process, particularly the first and second moments, which are called the mean, variance and autocovariance functions. These will now be defined for discrete time, with similar definitions applying in continuous time.

34

Mean   The mean function, $\mu(t)$, is defined by

$$\mu(t) = E(X_t).$$

Variance   The variance function, $\sigma^2(t)$, is defined by

$$\sigma^2(t) = \text{Var}(X_t).$$

Autocovariance   The autocovariance function, $\gamma(t_1, t_2)$, is defined by

$$\gamma(t_1, t_2) = \text{Cov}(X_{t_1}, X_{t_2})$$
$$= E\{[X_{t_1} - \mu(t_1)][X_{t_2} - \mu(t_2)]\}$$

The variance function is a special case of $\gamma(t_1, t_2)$ when $t_1 = t_2$. (Readers who are unfamiliar with 'covariance' should read Appendix III.)

Higher moments of a stochastic process may be defined in an obvious way, but are rarely used in practice, since a knowledge of the two functions $\mu(t)$ and $\gamma(t_1, t_2)$ is often adequate.

3.2 Stationary processes

An important class of stochastic processes are those which are *stationary*. A heuristic idea of stationarity was introduced in Section 2.2.

A time series is said to be *strictly stationary* if the joint distribution of $X(t_1), \ldots, X(t_n)$ is the same as the joint distribution of $X(t_1 + \tau), \ldots, X(t_n + \tau)$ for all $t_1, \ldots, t_n, \tau$. In other words shifting the time origin by an amount τ has no effect on the joint distributions which must therefore depend only on the intervals between $t_1, t_2, \ldots, t_n$. The above definition holds for any value of n. In particular, if $n = 1$, it implies that the distribution of $X(t)$ must be the same for all t so that

$$\mu(t) = \mu$$
$$\sigma^2(t) = \sigma^2$$

are both constants which do not depend on the value of t.

Furthermore if $n = 2$, the joint distribution of $X(t_1)$ and $X(t_2)$ depends only on $(t_2 - t_1)$ which is called the *lag*. Thus the autocovariance function, $\gamma(t_1, t_2)$, also depends only on $(t_2 - t_1)$ and may be written as $\gamma(\tau)$ where

$$\gamma(\tau) = E[(X(t) - \mu)(X(t + \tau) - \mu)]$$

is called the autocovariance coefficient at lag τ. In future 'autocovariance function' will be abbreviated to acv.f.

The size of the autocovariance coefficients depend on the units in which $X(t)$ is measured. Thus, for interpretative purposes, it is useful to standardize the acv.f. to produce a function called the *autocorrelation* function which is given by

$$\rho(\tau) = \gamma(\tau)/\gamma(0),$$

and which measures the autocorrelation between $X(t)$ and $X(t + \tau)$. Its empirical counterpart was introduced in Section 2.7. In future, 'autocorrelation function' will be abbreviated to ac.f.

Note that $\gamma(\tau)$, $\rho(\tau)$ are discrete if the time series is discrete and continuous if the time series is continuous.

At first sight it may seem surprising to suggest that there are processes for which the distribution of $X(t)$ should be the same for all t. However readers with some knowledge of stochastic processes will know that there are many processes $\{X(t)\}$ which have what is called an *equilibrium* distribution as $t \to \infty$, in which the probability distribution of $X(t)$ tends to a limit which does *not* depend on the initial conditions. Thus once such a process has been running for some time, the distribution of $X(t)$ will change very little. Indeed if the initial conditions are specified to be identical to the equilibrium distribution, the process is stationary in time and the equilibrium distribution is then the stationary distribution of the process. Of course the *conditional* distribution of $X(t_2)$ given that $X(t_1)$ has taken a particular value, say $x(t_1)$, may be quite different from the stationary distribution, but this is perfectly consistent with the idea of stationarity and is where a knowledge of the ac.f or acv.f. is appropriate.

3.2.1 Second-order stationarity

In practice it is often useful to define stationarity in a less restricted way than that described above. A process is called second-order stationary (or weakly stationary) if its mean is constant and its acv.f. depends only on the lag, so that

$$E[X(t)] = \mu$$

and

$$\text{Cov}[X(t), X(t + \tau)] = \gamma(\tau).$$

No assumptions are made about higher moments than those of second order. Note that the above definition implies that the variance, as well as the mean, is constant. Also note that both the variance and the mean must be finite.

This weaker definition of stationarity will be used from now on, as many of the properties of stationary processes depend only on the structure of the process as specified by its first and second moments (Bartlett, 1966; Chapter 6). One important class of processes where this is particularly true is the class of *normal* processes where the joint distribution of $X(t_1), \ldots, X(t_n)$ is multivariate normal for all $t_1, \ldots, t_n$. The multivariate normal distribution is completely characterized by its first and second moments, and hence by $\mu(t)$ and $\gamma(t_1, t_2)$, and so it follows that second-order stationarity implies strict stationarity for normal processes. However for processes which are very 'non-normal', μ and $\gamma(\tau)$ may not adequately describe the process.

3.3 The autocorrelation function

We have already noted in Section 2.7 that the sample autocorrelation coefficients of an observed time series are an important set of statistics for describing the time series. Similarly the (theoretical) autocorrelation function (ac.f.) of a stationary stochastic process is an important tool for assessing its properties. This section investigates the general properties of the ac.f.

Suppose a stationary stochastic process $X(t)$ has mean μ, variance σ^2, acv.f. $\gamma(\tau)$, and ac.f. $\rho(\tau)$. Then

$$\rho(\tau) = \gamma(\tau)/\gamma(0) = \gamma(\tau)/\sigma^2.$$

Note that $\rho(0) = 1$.

Property 1   The ac.f. is an *even* function of the lag in that

$$\rho(\tau) = \rho(-\tau).$$

This is easily demonstrated since $\rho(\tau) = \gamma(\tau)/\sigma^2$ and $X(t)$ is stationary so that

$$\begin{aligned}
\gamma(\tau) &= \text{Cov}[X(t), X(t + \tau)] \\
&= \text{Cov}[X(t - \tau), X(t)] \\
&= \gamma(-\tau).
\end{aligned}$$

Property 2 $|\rho(\tau)| \leqslant 1$. This can be proved by noting that

$$\text{Var}[\lambda_1 X(t) + \lambda_2 X(t + \tau)] \geqslant 0$$

for any constants λ_1, λ_2. This variance is equal to

$$\lambda_1^2 \text{ Var}[X(t)] + \lambda_2^2 \text{ Var}[X(t + \tau)] +$$
$$2\lambda_1 \lambda_2 \text{ Cov}[X(t), X(t + \tau)] = (\lambda_1^2 + \lambda_2^2)\sigma^2 + 2\lambda_1 \lambda_2 \gamma(\tau)$$

When $\lambda_1 = \lambda_2 = 1$, we find

$$\gamma(\tau) \geqslant -\sigma^2$$

so that $\rho(\tau) \geqslant -1$.

When $\lambda_1 = 1, \lambda_2 = -1$, we find

$$\sigma^2 \geqslant \gamma(\tau)$$

so that $\rho(\tau) \leqslant +1$.

Property 3   Non-uniqueness. For a particular acv.f. there corresponds only one possible stationary normal process, as we have already remarked in Section 3.2.1 that a stationary normal process is completely determined by its mean, variance and ac.f. However, it is always possible to find many non-normal processes with the same ac.f. and this creates a further difficulty in interpreting sample ac.f's. Jenkins and Watts (1968; p. 170) give an example of two different stochastic processes which have the same ac.f.

3.4 Some useful stochastic processes

This section describes several different types of stochastic process which are sometimes useful in setting up a model for a time series.

3.4.1 A purely random process

A discrete process $\{Z_t\}$ is called a purely random process if the random variables $\{Z_t\}$ are a sequence of mutually independent, identically distributed variables. From the definition it follows that the process has constant mean and variance and that

$$\gamma(k) = \text{Cov}(Z_t, Z_{t+k})$$
$$= 0 \quad \text{for} \quad k = \pm 1, \pm 2, \dots \, .$$

As the mean and acv.f. do not depend on time, the process is second-order stationary. In fact it is clear that the process is also strictly stationary. The ac.f. is given by

$$\rho(k) = \begin{cases} 1 & k = 0 \\ 0 & k = \pm 1, \pm 2, \dots \end{cases}$$

A purely random process is sometimes called *white noise*, particularly by engineers. Processes of this type are useful in many situations particularly as constituents of more complicated processes such as moving average processes (Section 3.4.3.).

There has been some controversy as to whether it is possible to have a continuous time purely random process. Such a process would have an ac.f. given by

$$\rho(k) = \begin{cases} 1 & k = 0 \\ 0 & k \neq 0 \end{cases}$$

which is a discontinuous function. It can be shown that such a process would have infinite variance, and is a physically unrealizable phenomenon. However, it turns out to be a useful mathematical idealization for some theoretical purposes, and also as an approximation to certain processes which occur in practice, such as voltage fluctuations in a conductor due to thermal noise, and the impulses acting on a particle suspended in fluid along a given axis to produce Brownian motion.

Cox and Miller (1968) point out that a process approximating to white noise may be obtained by considering either

(a) a process in continuous time with ac.f. $\rho(\tau) = e^{-\lambda|\tau|}$ and letting $\lambda \to \infty$, so that the ac.f. decays very quickly;

or

(b) a purely random process in discrete time at intervals Δt and letting $\Delta t \to 0$.

More generally we may regard continuous white noise as any signal made up of the superposition of a large number of independent effects of very brief duration, which appears to behave like continuous white noise when sampled even at quite small discrete intervals.

3.4.2 Random walk   Suppose that $\{Z_t\}$ is a discrete, purely random process with mean μ and variance σ_Z^2. A process $\{X_t\}$

is said to be a random walk if

$$X_t = X_{t-1} + Z_t. \tag{3.1}$$

The process is customarily started at zero when $t = 0$ so that

$$X_1 = Z_1$$

and

$$X_t = \sum_{i=1}^{t} Z_i.$$

Then we find $E(X_t) = t\mu$ and that $\text{Var}(X_t) = t\sigma_Z^2$.

As the mean and variance change with t, the process is non-stationary.

However it is interesting to note that the first differences of a random walk, given by

$$Z_t = X_t - X_{t-1}$$

form a purely random process, which is therefore stationary.

The best-known examples of time series which behave very much like random walks are share prices. When these are analysed, a model which often gives a good approximation to the data is

Share price on day t = share price on day $(t - 1)$
+ random error.

3.4.3 Moving average processes Suppose that $\{Z_t\}$ is a purely random process with mean zero and variance σ_Z^2. Then a process $\{X_t\}$ is said to be a moving average process of order m if

$$X_t = \beta_0 Z_t + \beta_1 Z_{t-1} + \ldots + \beta_m Z_{t-m} \tag{3.2}$$

where $\{\beta_i\}$ are constants. The Z's are usually scaled so that $\beta_0 = 1$.

We find immediately that

$$E(X_t) = 0$$

41

and $\text{Var}(X_t) = \sigma_Z^2 \sum_{i=0}^{m} \beta_i^2$

since the Z's are independent. We also have

$$\begin{aligned} \gamma(k) &= \text{Cov}\,(X_t, X_{t+k}) \\ &= \text{Cov}\,(\beta_0 Z_t + \ldots + \beta_m Z_{t-m}, \beta_0 Z_{t+k} \\ &\quad + \ldots + \beta_m Z_{t+k-m}) \\ &= \begin{cases} 0 & k > m \\ \sigma_Z^2 \sum_{i=0}^{m-k} \beta_i \beta_{i+k} & k = 0, 1, \ldots, m \\ \gamma(-k) & k < 0 \end{cases} \end{aligned}$$

since $\text{Cov}(Z_s, Z_t) = \begin{cases} \sigma_Z^2 & s = t \\ 0 & s \neq t \end{cases}$

As $\gamma(k)$ does not depend on t, and the mean is constant, the process is second-order stationary for all values of the $\{\beta_i\}$. Furthermore if the Z's are normally distributed, so are the X's, and we have a normal process completely determined by the mean and acv.f. The process is then strictly stationary.

The ac.f. of the moving average process is given by

$$\rho(k) = \begin{cases} 1 & k = 0 \\ \sum_{i=0}^{m-k} \beta_i \beta_{i+k} / \sum_{i=0}^{m} \beta_i^2 & k = 1, \ldots, m \\ 0 & k > m \\ \rho(-k) & k < 0 \end{cases}$$

In particular for the first-order moving average process

$$X_t = Z_t + \beta Z_{t-1}$$

we find

$$\rho(k) = \begin{cases} 1 & k = 0 \\ \beta/(1 + \beta^2) & k = \pm 1 \\ 0 & \text{otherwise} \end{cases}$$

In future 'moving average' will be abbreviated to MA.

Although no restrictions on the $\{\beta_i\}$ are required for a MA process to be stationary, Box and Jenkins (1970, p. 50) propose restrictions on the $\{\beta_i\}$ to ensure what they call 'invertibility'. This condition is discussed by Chatfield and Prothero (1973b), and arises in the following way. Consider the following first order MA processes:

A) $\quad X_t = Z_t + \theta Z_{t-1}$

B) $\quad X_t = Z_t + \dfrac{1}{\theta} Z_{t-1}.$

It can easily be shown that these different processes have exactly the same ac.f. Thus we cannot identify a MA process uniquely from a given ac.f. Now, if we express models A and B by putting Z_t in terms of $X_t, X_{t-1}, \ldots$, we find by successive substitution that

A) $\quad Z_t = X_t - \theta X_{t-1} + \theta^2 X_{t-2} - \ldots$

B) $\quad Z_t = X_t - \dfrac{1}{\theta} X_{t-1} + \dfrac{1}{\theta^2} X_{t-2} - \ldots.$

If $|\theta| < 1$, the series for A converges whereas that for B does not. Thus an estimation procedure which involves estimating the residuals (see Section 4.3.1) will lead naturally to model A. Thus if $|\theta| < 1$, model A is said to be invertible whereas model B is not. The imposition of the invertibility condition ensures that there is a unique MA process for a given ac.f.

The invertibility condition for the general order MA process is best expressed by using the backward shift operator, denoted by B, which is defined by

$$B^j X_t = X_{t-j} \qquad \text{for all } j.$$

Then equation (3.2) may be written as

$$X_t = (\beta_0 + \beta_1 B + \ldots + \beta_m B^m) Z_t$$
$$= \theta(B) Z_t$$

where $\theta(B)$ is a polynomial of order m in B. A MA process of order m is invertible if the roots of the equation (regarding B as a complex variable and not an operator)

$$\theta(B) = \beta_0 + \beta_1 B + \ldots + \beta_m B^m = 0$$

all lie outside the unit circle (Box and Jenkins, 1970; p. 50).

MA processes have been used in many areas, particularly econometrics. For example economic indicators are affected by a variety of 'random' events such as strikes, government decisions, shortage of key materials and so on. Such events will not only have an immediate effect but may also affect economic indicators to a lesser extent in several subsequent periods, and so it is at least plausible that a MA process may be appropriate.

Recently Box and Jenkins (1970) have advocated the use of MA processes in conjunction with processes called autoregressive processes which are considered in the next section.

Note that an arbitrary constant, μ say, may be added to the right hand side of equation (3.2) to give a process with mean μ. This does not affect the ac.f. and has been omitted for simplicity.

3.4.4 Autoregressive processes   Suppose that $\{Z_t\}$ is a purely random process with mean zero and variance σ_Z^2. Then a process $\{X_t\}$ is said to be an autoregressive process of order m if

$$X_t = \alpha_1 X_{t-1} + \ldots + \alpha_m X_{t-m} + Z_t. \tag{3.3}$$

This is rather like a multiple regression model, but X_t is regressed not on independent variables but on past values of X_t. Hence the term *auto*regressive. In future 'autoregressive' will be abbreviated to AR. Processes of this type were introduced by G. U. Yule in about 1920.

(a) First-order process   For simplicity, we begin by examining the first-order case, where $m = 1$. Then

$$X_t = \alpha X_{t-1} + Z_t \tag{3.4}$$

The first-order AR process is sometimes called the Markov process, after the Russian A. A. Markov. By successive substitution in (3.4) we may write

$$X_t = \alpha[\alpha X_{t-2} + Z_{t-1}] + Z_t$$
$$= \alpha^2[\alpha X_{t-3} + Z_{t-2}] + \alpha Z_{t-1} + Z_t$$

and eventually we find that X_t may be expressed as an infinite-order MA process in the form

$$X_t = Z_t + \alpha Z_{t-1} + \alpha^2 Z_{t-2} + \ldots$$

This duality between AR and MA processes is useful for a variety of purposes. Rather than use successive substitution, it is simpler to use the backward shift operator B. Then equation (3.4) may be written

$$(1 - \alpha B)X_t = Z_t$$

so that

$$X_t = Z_t/(1 - \alpha B)$$
$$= (1 + \alpha B + \alpha^2 B^2 + \ldots)Z_t$$
$$= Z_t + \alpha Z_{t-1} + \alpha^2 Z_{t-2} + \ldots .$$

Then we find

$$E(X_t) = 0$$

and

$$\mathrm{Var}(X_t) = \sigma_Z^2(1 + \alpha^2 + \alpha^4 + \ldots)$$

Thus the variance is finite provided that $|\alpha| < 1$, in which case

$$\mathrm{Var}(X_t) = \sigma_Z^2/(1 - \alpha^2) = \sigma_X^2.$$

The acv.f. is given by

$$\gamma(k) = E[X_t X_{t+k}]$$
$$= E\{[\Sigma\alpha^i Z_{t-i}][\Sigma\alpha^j Z_{t+k-j}]\}$$
$$= \sigma_Z^2 \sum_{i=0}^{\infty} \alpha^i \alpha^{k+i} \qquad \text{(for } k \geqslant 0)$$

which converges for $|\alpha| < 1$ to

$$\gamma(k) = \alpha^k \sigma_Z^2/(1 - \alpha^2)$$
$$= \alpha^k \sigma_X^2$$

For $k < 0$, we find $\gamma(k) = \gamma(-k)$. Since $\gamma(k)$ does not depend on t, an AR process of order 1 is second-order stationary provided that $|\alpha| < 1$. The ac.f. is given by

$$\rho(k) = \alpha^k \qquad (k = 0, 1, 2, \ldots)$$

To get an even function defined for all integer k we take

$$\rho(k) = \alpha^{|k|} \qquad (k = 0, \pm 1, \pm 2, \ldots)$$

The ac.f. may in fact be obtained more simply in the following way, by assuming *a priori* that the process is stationary in which case $E(X_t) = 0$. Multiply through equation (3.4) by X_{t-k} (NOT X_{t+k}!) and take expectations. Then we find, for $k > 0$, that

$$\gamma(-k) = \alpha\gamma(-k + 1)$$

since $E(Z_t X_{t-k}) = 0$ for $k > 0$. Since $\gamma(k)$ is an even function, we must also have

$$\gamma(k) = \alpha\gamma(k - 1) \qquad \text{for } k > 0.$$

Now $\gamma(0) = \sigma_X^2$, and so $\gamma(k) = \alpha^k \sigma_X^2$ for $k \geqslant 0$. Thus $\rho(k) = \alpha^k$ for $k \geqslant 0$. Now since $|\rho(k)| \leqslant 1$, we must have $|\alpha| \leqslant 1$. But if $|\alpha| = 1$, then $|\rho(k)| = 1$ for all k, which is a degenerate case. Thus $|\alpha| < 1$ for a proper stationary process.

The above method of obtaining the ac.f. is often used,

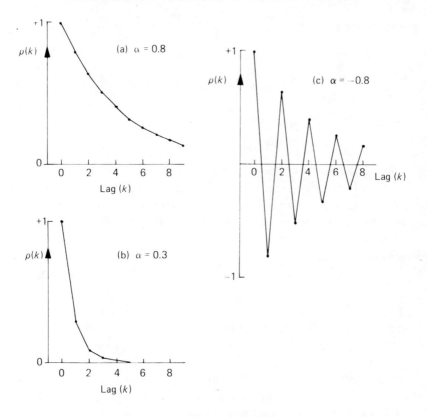

Figure 3.1   Three examples of the autocorrelation function of a first-order autoregressive process with (a) $\alpha = 0.8$; (b) $\alpha = 0.3$; (c) $\alpha = -0.8$.

even though it involves an initial assumption of stationarity.

Three examples of the ac.f. of a first-order AR process are shown in Fig. 3.1 for $\alpha = 0.8$, 0.3, −0.8. Note how quickly the ac.f. decays when $\alpha = 0.3$, and note how the ac.f. alternates when α is negative.

(b) General-order case   As in the first-order case, we can express an AR process of finite order as a MA process of infinite order. This may be done by successive substitution, or by using the backward shift operator. Then equation 3.3

may be written as

$$(1 - \alpha_1 B - \ldots - \alpha_m B^m)X_t = Z_t$$

or $\quad X_t = Z_t/(1 - \alpha_1 B - \ldots - \alpha_m B^m)$

$$= f(B) Z_t$$

where $f(B) = (1 - \alpha_1 B - \ldots - \alpha_m B^m)^{-1}$

$$= (1 + \beta_1 B + \beta_2 B^2 + \ldots)$$

The relationship between the α's and the β's may then be found. Having expressed X_t as a MA process, it follows that $E(X_t) = 0$. The variance is finite provided that $\Sigma \beta_i^2$ converges, and this is a necessary condition for stationarity. The acv.f. is given by

$$\gamma(k) = \sigma_Z^2 \sum_{i=0}^{\infty} \beta_i \beta_{i+k} \qquad (\text{where } \beta_0 = 1)$$

A sufficient condition for this to converge, and hence for stationarity, is that $\Sigma \mid \beta_i \mid$ converges.

We can in principle find the ac.f. of the general-order AR process using the above procedure, but the $\{\beta_i\}$ may be algebraically hard to find. The alternative simpler way is to *assume* the process is stationary, multiply through equation (3.3) by X_{t-k}, take expectations, and divide by σ_X^2 assuming that the variance of X_t is finite. Then, using the fact that $\rho(k) = \rho(-k)$ for all k, we find

$$\rho(k) = \alpha_1 \rho(k - 1) + \ldots + \alpha_m \rho(k - m) \qquad \text{for all}$$
$$k > 0.$$

This set of equations is called the Yule–Walker equations after G. U. Yule and Sir Gilbert Walker. They are a set of difference equations and have the general solution

$$\rho(k) = A_1 \pi_1^{|k|} + \ldots + A_m \pi_m^{|k|}$$

where $\{\pi_i\}$ are the roots of

$$y^m - \alpha_1 y^{m-1} \ldots - \alpha_m = 0.$$

The constants $\{A_i\}$ are chosen to satisfy the initial conditions depending on $\rho(0) = 1$. Thus we have $\Sigma A_i = 1$, and the first $(m - 1)$ Yule–Walker equations provide $(m - 1)$ further restrictions on the $\{A_i\}$ using $\rho(0) = 1$ and $\rho(k) = \rho(-k)$.

From the general form of $\rho(k)$, it is clear that $\rho(k)$ tends to zero as k increases provided that $|\pi_i| < 1$ for all i, and this is a necessary and sufficient condition for the process to be stationary.

An equivalent way of expressing the stationarity condition is to say that the roots of the equation

$$\phi(B) = 1 - \alpha_1 B - \ldots - \alpha_m B^m = 0$$

must lie outside the unit circle (Box and Jenkins, 1970; Section 3.2).

Of particular interest is the second-order AR process when $m = 2$. Then π_1, π_2 are the roots of the quadratic equation

$$y^2 - \alpha_1 y - \alpha_2 = 0.$$

Thus $|\pi_i| < 1$ if

$$\left| \frac{\alpha_1 \pm \sqrt{(\alpha_1^2 + 4\alpha_2)}}{2} \right| < 1$$

from which it can be shown, (exercise 6), that the stationarity region is the triangular region satisfying

$$\left. \begin{array}{l} \alpha_1 + \alpha_2 < 1 \\ \alpha_1 - \alpha_2 > -1 \\ \alpha_2 > -1 \end{array} \right\}$$

The roots are real if $\alpha_1^2 + 4\alpha_2 > 0$, in which case the ac.f. decreases exponentially with k, but the roots are complex if $\alpha_1^2 + 4\alpha_2 < 0$ in which case we find that the ac.f. is a damped cosine wave. (See for example Cox and Miller, 1968; Section 7.2).

When the roots are real, the constants A_1, A_2 are found as follows. Since $\rho(0) = 1$, we have

$$A_1 + A_2 = 1.$$

From the first of the Yule–Walker equations, we have

$$\rho(1) = \alpha_1 \rho(0) + \alpha_2 \rho(-1)$$
$$= \alpha_1 + \alpha_2 \rho(1)$$

Thus $\rho(1) = \alpha_1 /(1 - \alpha_2)$

$$= A_1 \pi_1 + A_2 \pi_2$$
$$= A_1 \pi_1 + (1 - A_1)\pi_2.$$

Hence we find

$$A_1 = [\alpha_1 /(1 - \alpha_2) - \pi_2]/(\pi_1 - \pi_2)$$

and $A_2 = 1 - A_1$.

AR processes have been applied to many situations in which it is reasonable to assume that the present value of a time series depends on the immediate past values together with a random error. For simplicity we have only considered processes with mean zero, but non-zero means may be dealt with by rewriting equation 3.3 in the form

$$X_t - \mu = \alpha_1 (X_{t-1} - \mu) + \ldots + \alpha_m (X_{t-m} - \mu) + Z_t.$$

This does not affect the ac.f.

3.4.5 Mixed models A useful class of models for time series is that formed from a combination of MA and AR processes. A mixed autoregressive-moving average process (abbreviated ARMA process) which contains p AR terms and q MA terms is said to be of order (p, q) and is given by

$$\begin{aligned} X_t = \alpha_1 X_{t-1} + \ldots + \alpha_p X_{t-p} + Z_t + \beta_1 Z_{t-1} \\ + \ldots + \beta_q Z_{t-q} \end{aligned} \quad (3.5)$$

Using the backward shift operator B, equation (3.5) may be written in the form

$$\phi(B)X_t = \theta(B)Z_t \qquad (3.5a)$$

where $\phi(B)$, $\theta(B)$ are polynomials of order p, q respectively, such that

$$\phi(B) = 1 - \alpha_1 B - \ldots - \alpha_p B^p$$

and

$$\theta(B) = 1 + \beta_1 B + \ldots + \beta_q B^q$$

The values of α_i which make the process stationary are such that the roots of

$$\phi(B) = 0$$

lie outside the unit circle. The values of $\{\beta_i\}$ which make the process invertible are such that the roots of

$$\theta(B) = 0$$

lie outside the unit circle.

It is fairly straightforward, though algebraically rather tedious, to calculate the ac.f. of an ARMA process, but this will not be discussed here. (See exercise 11 and Box and Jenkins, 1970; Section 3.4).

The importance of ARMA processes lies in the fact that a stationary time series may often be described by an ARMA model involving fewer parameters than a MA or AR process by itself.

3.4.6 Integrated models   In practice most time series are non-stationary. In order to fit a stationary model, such as those discussed in Sections 3.4.3–3.4.5, it is necessary to remove non-stationary sources of variation. If the observed time series is non-stationary in the mean then we can difference the series, as suggested in Section 2.5.3, and this approach is widely used in econometrics. If X_t is replaced by $\nabla^d X_t$ in equation 3.5 then we have a model capable of describing certain types of non-stationary series. Such a model is called an 'integrated' model because the stationary model which is fitted to the differenced data has to be summed or 'integrated' to provide a model for the non-stationary data. Writing

$$W_t = \nabla^d X_t,$$

the general autoregressive integrated moving average process (abbreviated ARIMA process) is of the form

$$W_t = \alpha_1 W_{t-1} + \ldots + \alpha_p W_{t-p} + Z_t + \ldots + \beta_q Z_{t-q} \tag{3.6}$$

It is interesting to note that the random walk is a particular example of an ARIMA process, when written as

$$\nabla X_t = Z_t.$$

3.4.7 The general linear process The infinite-order MA process with non-zero mean given by

$$X_t - \mu = \sum_{i=0}^{\infty} \beta_i Z_{t-i} \tag{3.7}$$

is often called the general linear process since processes of this type can be obtained by passing a purely random process through a linear system (see Chapter 9). Both MA and AR processes are special cases of the general linear process and the duality between these two classes is easily demonstrated using the backward shift operator. A MA process of finite order can be expressed as an AR process of infinite order, while an AR process of finite order can be expressed as a MA process of infinite order (see exercises 9 and 10).

3.4.8 Continuous processes So far, we have only considered stochastic processes in discrete time, because these are the main type of process the statistician uses in practice. Continuous processes have been used in some applications, notably in the study of control theory by electrical engineers. Here we shall only indicate some of the problems connected with their use.

By analogy with equation (3.7), we could try to define the general linear process in continuous time by

$$X(t) = \int_0^\infty h(u)\, Z(t - u)\, du, \qquad (3.8)$$

but if $Z(t)$ is continuous white noise, there are immediate mathematical difficulties regarding the above integral (Cox and Miller, 1968; Section 7.4). As discussed in Section 3.4.1, continuous white noise is not physically realizable and so equation (3.8) is not generally used. Most authors, such as Yaglom (1962; p. 22), only consider continuous processes which have a continuous ac.f. However Jenkins and Watts (1968) do use equation (3.8).

As an example of a continuous process, we will briefly consider a first-order continuous AR process. A first-order discrete AR process (equation 3.4) may be written in terms of X_t, ∇X_t and Z_t. As differencing in discrete time corresponds to differentiation in continuous time, a natural way of defining a first-order AR process is by

$$\frac{dX(t)}{dt} + aX(t) = Z(t) \qquad (3.9)$$

which is essentially the Langevin equation (Cox and Miller, 1968; p. 298). Yaglom (1962; p. 69) suggests that it is more legitimate to write equation (3.9) in the form

$$dX(t) + aX(t)dt = dU(t) \qquad (3.10)$$

where $\{U(t)\}$ is a process with orthogonal increments (i.e. the random variables $[U(t_2) - U(t_1)]$ and $[U(t_4) - U(t_3)]$ are independent for any two non-over-lapping intervals

(t_1, t_2) and (t_3, t_4)). Equation (3.10) is the same as that which defines the Ornstein–Uhlenbeck process (Cox and Miller, 1968; Section 5.8). It can be shown that the process $X(t)$ defined by equation (3.10) has ac.f.

$$\rho(u) = e^{-a|u|}$$

which is similar to the ac.f. of a first-order discrete AR process in that both decay exponentially.

The rigorous study of continuous processes requires considerable mathematical skill, including a knowledge of stochastic integration, and we will not pursue it here.

3.5 The Wold decomposition theorem

This section gives a brief introduction to a famous theorem given by H. Wold in 1938 which applies to any stationary process in discrete time. The treatment in this section is a shortened version of that given by Cox and Miller (1968). The Wold decomposition theorem says that any discrete stationary process can be expressed as the sum of two uncorrelated processes, one purely deterministic and one which is purely indeterministic. The terms deterministic and indeterministic are defined as follows. We can regress X_t on $(X_{t-q}, X_{t-q-1} \ldots)$ and denote the residual variance from the resulting linear regression model by τ_q^2. As $\tau_q^2 \leqslant \mathrm{Var}(X_t)$, it is clear that, as q increases, τ_q^2 is a non-decreasing bounded sequence and therefore tends to a limit as $q \to \infty$. If this limit is equal to the variance of X_t, then linear regression on the remote past is useless for prediction purposes and we say that $\{X_t\}$ is *purely indeterministic*. However if $\lim_{q \to \infty} \tau_q^2$ is less than the variance of X_t, then we say that X_t contains a purely deterministic component.

All the stationary processes we have considered in this chapter, such as AR and MA processes, are purely indeterministic. The best-known examples of purely deterministic processes are sinusoidal processes (see exercise 14) such as

$$X_t = g \cos (\omega t + \theta) \qquad (3.11)$$

where g is a constant, ω is a constant in $(0, \pi)$ called the frequency of the process and θ is a random variable called the phase which is uniformly distributed on $(0, 2\pi)$ but which is fixed for a single realization. Note that we must include the term θ so that

$$E(X_t) = 0 \qquad \text{for all } t,$$

otherwise (3.11) would not define a stationary process. As θ is fixed for a single realization, once enough values of X_t have been observed to evaluate θ, all subsequent values of X_t are completely determined. It is then obvious that (3.11) defines a purely deterministic process.

The Wold decomposition theorem also says that the purely indeterministic component can be written as the linear sum of an 'innovation' process $\{Z_t\}$, which is a sequence of uncorrelated random variables. A special class of processes of particular interest arise when the Z's are independent and not merely uncorrelated, as we then have a general linear process (Section 3.4.7). On the other hand when processes are generated in a non-linear way the Wold decomposition is usually of little interest.

The concept of a purely indeterministic process is a useful one and most of the stationary stochastic processes which are considered in the rest of this book are of this type.

Exercises

In all the following questions $\{Z_t\}$ is a discrete, purely random process, such that $E(Z_t) = 0$, $V(Z_t) = \sigma_Z^2$, $\text{Cov}(Z_t, Z_{t+k}) = 0$, $k \neq 0$.

1. Find the ac.f. of the second-order MA process given by

$$X_t = Z_t + 0.7Z_{t-1} - 0.2Z_{t-2}.$$

2. Show that the ac.f. of the mth-order MA process given by

$$X_t = \sum_{t=0}^{m} Z_t/(m+1)$$

is

$$\rho(k) = \begin{cases} (m+1-k)/(m+1) & k = 0, 1, \ldots, m. \\ 0 & k > m \end{cases}$$

3. Show that the infinite-order MA process $\{X_t\}$ defined by

$$X_t = Z_t + C(Z_{t-1} + Z_{t-2} + \ldots)$$

where C is a constant, is non-stationary.
Also show that the series of first differences $\{Y_t\}$ defined by

$$Y_t = X_t - X_{t-1}$$

is a first-order MA process and is stationary. Find the ac.f. of $\{Y_t\}$

4. Find the ac.f. of the first-order AR process defined by

$$X_t = 0.7X_{t-1} + Z_t.$$

Plot $\rho(k)$ for $k = -6, -5, \ldots, -1, 0, +1, \ldots, +6$.

5. If $X_t = \mu + Z_t + \beta Z_{t-1}$, where μ is a constant, show that the ac.f. does not depend on μ.

6. Find the values of λ_1, λ_2, such that the second-order AR process defined by

$$X_t = \lambda_1 X_{t-1} + \lambda_2 X_{t-2} + Z_t$$

is stationary.

If $\lambda_1 = \frac{1}{3}$, $\lambda_2 = \frac{2}{9}$, show that the ac.f. of X_t is given by

$$\rho(k) = \frac{16}{21} \left(\frac{2}{3}\right)^{|k|} + \frac{5}{21} \left(-\frac{1}{3}\right)^{|k|}.$$

$$k = 0, \pm 1, \pm 2, \ldots$$

7. Explain what is meant by a weakly (or second-order) stationary process, and define the ac.f. $\rho(u)$ for such a process. Show that $\rho(u) = \rho(-u)$ and that $|\rho(u)| \leqslant 1$.

   Show that the ac.f. of the stationary second-order AR process,

$$X_t = \frac{1}{12}X_{t-1} + \frac{1}{12}X_{t-2} + Z_t$$

is given by $\rho(k) = \dfrac{45}{77} \left(\dfrac{1}{3}\right)^{|k|} + \dfrac{32}{77} \left(-\dfrac{1}{4}\right)^{|k|}.$

$$k = 0, \pm 1, \pm 2, \ldots$$

8. The stationary process $\{X_t\}$ has acv.f. $\gamma_X(k)$. A new stationary process $\{Y_t\}$ is defined by $Y_t = X_t - X_{t-1}$. Obtain the acv.f. of $\{Y_t\}$ in terms of $\gamma_X(k)$ and find $\gamma_Y(k)$ when $\gamma_X(k) = \lambda^{|k|}$.

9. Find the MA representation of the first-order AR process

$$X_t = 0.3X_{t-1} + Z_t.$$

10. Suppose that a stationary process $\{X_t\}$ can be represented in two equivalent forms by

$$a_0 X_t + a_1 X_{t-1} + \ldots = Z_t$$

and

$$X_t = b_0 Z_t + b_1 Z_{t-1} + \ldots$$

where we may take $a_0 = b_0 = 1$. These are the AR and

MA representations of the process. Let

$$A(s) = \sum_{i=0}^{\infty} a_i s^i$$

and $B(s) = \sum_{i=0}^{\infty} b_i s^i$. Show that $A(s) B(s) = 1$.

By defining $\Gamma(s) = \sum_{k=-\infty}^{\infty} \gamma_X(k) s^k$ also show that

$$\Gamma(s) = \sigma_Z^2 \, B(s) \, B\left(\frac{1}{s}\right) = \sigma_Z^2 \, / \left[A(s) \, A\left(\frac{1}{s}\right) \right].$$

11. Show that the ac.f. of the mixed ARMA model

$$X_t = \alpha X_{t-1} + Z_t + \beta Z_{t-1}$$

is given by

$$\rho(1) = (1 + \alpha\beta)(\alpha + \beta)/(1 + \beta^2 + 2\alpha\beta)$$
$$\rho(k) = \alpha\rho(k-1) \qquad (k = 2, 3, \ldots)$$

12. A Markov process $X(t)$ in continuous time has two possible states, 0 and 1. Let

$$P[X(t + \Delta t) = 1 \mid X(t) = 0] = \alpha\Delta t + o(\Delta t)$$
$$P[X(t + \Delta t) = 0 \mid X(t) = 1] = \beta\Delta t + o(\Delta t)$$

It can be shown that

$$P[X(t) = 0] = \beta/(\alpha + \beta) + [P(X(0) = 0) \\ - \beta/(\alpha + \beta)] \, e^{-(\alpha+\beta)t}.$$

If $P(X(0) = 0) = \beta/(\alpha + \beta)$ show that the process is second-order stationary.
(Hint: Show that $E(X(t) X(t + \tau)) =$

$$\frac{\alpha}{\alpha + \beta} [\alpha/(\alpha + \beta) + (1 - \alpha/(\alpha + \beta)) \, e^{-(\alpha+\beta)\tau}]$$

is a function of τ only.)

13. For a continuous stationary process $X(t)$ define

$$g(T) = \mathrm{Var}\left[\frac{1}{T}\int_0^T X(t)\mathrm{d}t\right].$$

Prove that

$$g(T) = \frac{2}{T^2}\int_0^T (T-h)\gamma_X(h)\,\mathrm{d}h$$

and

$$\gamma_X(h) = \frac{1}{2}\frac{\mathrm{d}^2}{\mathrm{d}h^2}[h^2 g(h)]$$

14. For a complex-valued process $X(t)$, with (complex) mean μ, the acv.f. is defined by

$$\gamma(\tau) = E[(X(t) - \mu)(\overline{X}(t+\tau) - \bar{\mu})]$$

where the overbar denotes the complex conjugate. Show that the process $X(t) = Ye^{i\omega t}$ is second-order stationary where Y is a complex random variable mean zero which does not depend on t, and ω is a real constant. One useful form for the random variable Y occurs when it takes the form $ge^{i\theta}$ where g is a constant and θ is a uniformly distributed random variable on $(0, 2\pi)$. Show that $E(Y) = 0$ in this case (see Yaglom, 1962; Section 2.8, but note that the autocovariance function is called the correlation function by Yaglom).

CHAPTER 4

Estimation in the Time Domain

In Chapter 3 we introduced several different types of
probability model which may be used to describe time series. In
this chapter we discuss the problem of fitting a suitable
model to an observed time series, confining ourselves to the
discrete case. The major diagnostic tool in this chapter is the
sample autocorrelation function. Inference based on this
function is often called analysis in the *time domain.*

4.1 Estimating the autocovariance and autocorrelation functions

We have already noted in Section 3.3 that the theoretical
ac.f. is an important tool for describing the properties of a
stationary stochastic process. In Section 2.7, we heuristically
introduced the sample ac.f. of an observed time series, and
this is an intuitively reasonable estimate of the theoretical
ac.f., provided the series is stationary. This section
investigates the properties of the sample ac.f. more closely.

Let us look first at the autocovariance function (acv.f.).
The sample autocovariance coefficient at lag k (see equation
(2.6)) given by

$$c_k = \sum_{t=1}^{N-k} (x_t - \bar{x})(x_{t+k} - \bar{x})/N \qquad (4.1)$$

is the usual estimator for the theoretical autocovariance
coefficient at lag k, $\gamma(k)$. The properties of this estimator are

discussed by Jenkins and Watts (1968, Section 5.3.3). It can be shown that

$$E(c_k) = (1 - k/N)[\gamma(k) - \text{Var}(\bar{x})],$$

so that the estimator is biased. However

$$\lim_{N \to \infty} E(c_k) = \gamma(k),$$

so that the estimator is asymptotically unbiased.

It can also be shown that

$$\text{Cov}(c_k, c_m) \simeq \sum_{r=-\infty}^{\infty} \{\gamma(r)\gamma(r + m - k) + \gamma(r + m)\gamma(r - k)\}/N \quad (4.2)$$

When $m = k$, formula (4.2) gives us the variance of c_k and hence the mean square error of c_k. Formula (4.2) also highlights the fact that successive values of c_k may be highly correlated and this increases the difficulty of interpreting the correlogram.

Jenkins and Watts (1968, Chapter 5) compare the estimator (4.1) with the alternative estimator

$$c_k' = \sum_{t=1}^{N-k} (x_t - \bar{x})(x_{t+k} - \bar{x})/(N - k).$$

This is used by some authors because it has a smaller bias, but Jenkins and Watts conjecture that it generally has a higher mean square error.

A third method of estimating the acv.f. is to use Quenouille's method of bias reduction, otherwise known as *jackknife* estimation. In this procedure the time series is divided into two halves, and the sample acv.f. is estimated from each half of the series and also from the whole series. If the three resulting estimates of $\gamma(k)$ are denoted by c_{k1}, c_{k2} and c_k in an obvious notation, then the jackknife estimate is given by

$$\tilde{c}_k = 2c_k - \tfrac{1}{2}(c_{k1} + c_{k2}) \quad (4.3)$$

It can be shown that this estimator reduces the bias from order $1/N$ to order $1/N^2$. It has an extra advantage in that one can see if both halves of the time series have similar properties and hence see if the time series is stationary. However, the method has the disadvantage that it requires extra computation. It is also sensitive to non-stationarity in the mean and c_{k1}, c_{k2}, should be compared with the overall c_k as well as with each other.

Having estimated the acv.f., we then take

$$r_k = c_k/c_0 \qquad\qquad (4.4)$$

as an estimator for $\rho(k)$. The properties of r_k are rather more difficult to find than those of c_k because it is the ratio of two random variables. It can be shown that r_k is generally biased. The bias can be reduced by jackknifing as in equation (4.3). The jackknife estimator is given in an obvious notation by

$$\tilde{r}_k = 2r_k - \tfrac{1}{2}(r_{k1} + r_{k2})$$

A general formula for the variance of r_k is given by Kendall and Stuart (1966) and depends on *all* the autocorrelation coefficients of the process. We will only consider the properties of r_k when sampling from a purely random process, when all the theoretical autocorrelation coefficients are zero except at lag zero. These results are the most useful since they enable us to decide if the observed values of r_k from a given time series are significantly different from zero.

Suppose that $x_1, \ldots, x_N$ are independent and identically distributed random variables with arbitrary mean. Then it can be shown (Kendall and Stuart, 1966, Chapter 48) that

$$E(r_k) \simeq -1/N$$

and

$$\mathrm{Var}(r_k) \simeq 1/N$$

and that r_k is asymptotically normally distributed under weak conditions. Thus having plotted the correlogram, as described in Section 2.7, we can plot approximate 95% confidence limits at $-1/N \pm 2/\sqrt{N}$, which are often further approximated to $\pm 2/\sqrt{N}$. Observed values of r_k which fall outside these limits are 'significantly' different from zero at the 5% level. However, when interpreting a correlogram, it must be remembered that the overall probability of getting a coefficient outside these limits, given that the data really are random, increases with the number of coefficients plotted. For example if the first 20 values of r_k are plotted then one expects one 'significant' value even if the data are random. Thus, if only one or two coefficients are 'significant', the size and lag of these coefficients must be taken into account when deciding if a set of data is random. Values well outside the 'null' confidence limits indicate non-randomness. So also do significant coefficients at a lag which has some physical interpretation, such as lag one or a lag corresponding to seasonal variation.

Fig. 4.1 shows the correlogram for 100 observations, generated on a computer, which are supposed to be independent normally distributed variables. The confidence limits are approximately $\pm 2/\sqrt{100} = \pm 0.2$. We see that 2 of the first 20 values of r_k lie just outside the significance limits. As these occur at apparently arbitrary lags (namely 12 and 17) we conclude that there is no firm evidence to disprove the hypothesis that the observations are independently distributed.

4.1.1 Interpreting the correlogram   We have already given some general advice on interpreting the correlogram in Section 2.7.2. For stationary series, the correlogram can be used to indicate if a MA, AR, or mixed model is appropriate and also to indicate the order of such a process. For example, suppose we find that r_1 is significantly different from zero but that subsequent values of r_k are all close to zero. Then

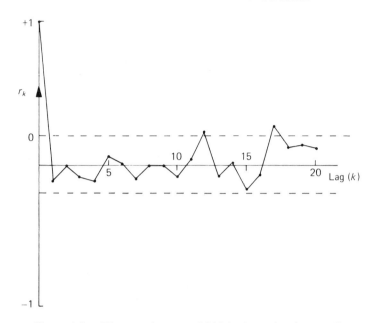

Figure 4.1   The correlogram of 100 'independent' normally distributed observations.

this would appear to indicate that a first-order MA process is appropriate since its theoretical ac.f. is of the same form as the sample ac.f.* Alternatively if r_1, r_2, r_3 appear to be decreasing geometrically, then a first-order AR process may be appropriate. If r_1 and r_2 are 'large' but other values of r_k are 'small', then two MA terms are indicated. More generally one has to work out the theoretical ac.f. of different ARMA models, compare them with the observed ac.f. and choose the model which appears to be most appropriate. In most cases ARMA models of low order are adequate.

The interpretation of correlograms is one of the hardest aspects of time-series analysis and considerable experience is

---

*The partial autocorrelation function (see Section 4.2.2) should also be examined, as it is possible, for example, for a second-order AR process to have an ac.f. of similar type.

required. It is a crucial part of the procedure suggested by Box and Jenkins (1970) and these authors give the theoretical ac.f's. of most low-order ARMA models so that the model which best agrees with the sample ac.f. may be chosen.

4.1.2 Ergodic theorems The fact that one can obtain consistent estimates of the properties of a stationary process from a single finite realization is not immediately obvious. However some theorems, called ergodic theorems, have been proved which show that, for most stationary processes which are likely to be met in practice, the sample moments of an observed record of length T converge (in mean square) to the corresponding population moments as $T \to \infty$. In other words time averages for a single realization converge to ensemble averages. See for example Yaglom (1962, Section 1.4). We will not pursue the topic here.

4.2 Fitting an autoregressive process

Having estimated the ac.f. of a given time series, we should have some idea as to what stochastic process will provide a suitable model. If an AR process is thought to be appropriate, there are two related questions:

(a) What is the order of the process?
(b) How can we estimate the parameters of the process?

We will consider question (b) first.

4.2.1. Estimating the parameters of an autoregressive process

Suppose we have an AR process of order m, with mean μ, given by

$$X_t - \mu = \alpha_1 (X_{t-1} - \mu) + \ldots + \alpha_m (X_{t-m} - \mu) + Z_t$$

$$(4.5)$$

65

Given N observations, $x_1, \ldots, x_N$, the parameters μ, $\alpha_1, \ldots, \alpha_m$, may be estimated by least squares by minimising

$$S = \sum_{t=m+1}^{N} [x_t - \mu - \alpha_1(x_{t-1} - \mu) - \ldots$$
$$- \alpha_m(x_{t-m} - \mu)]^2$$

with respect to $\mu, \alpha_1, \ldots, \alpha_m$. If the Z_t process is normal, then the least squares estimates are in addition maximum likelihood estimates (Jenkins and Watts, 1968, Section 5.4) conditional on the first m values in the time series being fixed.

In the first-order case, with $m = 1$, we find (see exercise 1).

$$\hat{\mu} = \frac{\bar{x}_{(2)} - \hat{\alpha}_1 \bar{x}_{(1)}}{1 - \hat{\alpha}_1} \tag{4.6}$$

and

$$\hat{\alpha}_1 = \frac{\sum\limits_{t=1}^{N-1} (x_t - \hat{\mu})(x_{t+1} - \hat{\mu})}{\sum\limits_{t=1}^{N-1} (x_t - \hat{\mu})^2} \tag{4.7}$$

where $\bar{x}_{(1)}, \bar{x}_{(2)}$ are the means of the first and last $(N-1)$ observations. Now since

$$\bar{x}_{(1)} \simeq \bar{x}_{(2)} \simeq \bar{x}$$

we have approximately that

$$\hat{\mu} \simeq \bar{x}. \tag{4.8}$$

This approximate estimator, which is intuitively appealing, is nearly always used in preference to (4.6). Substituting this value into (4.7) we have

$$\hat{\alpha}_1 = \frac{\sum\limits_{t=1}^{N-1} (x_t - \bar{x})(x_{t+1} - \bar{x})}{\sum\limits_{t=1}^{N-1} (x_t - \bar{x})^2}. \tag{4.9}$$

It is interesting to note that this is exactly the same estimator that would arise if we were to treat the autoregressive equation

$$X_t - \bar{x} = \alpha_1 (x_{t-1} - \bar{x}) + Z_t$$

as an ordinary regression with $(x_{t-1} - \bar{x})$ as the 'independent' variable. In fact H. B. Mann and A. Wald showed in 1943 that, asymptotically, much of classical regression theory can be applied to autoregressive situations.

A further approximation which is often used is obtained by noting that the denominator of (4.9) is approximately

$$\sum_{t=1}^{N} (x_t - \bar{x})^2,$$

so that

$$\hat{\alpha}_1 \simeq c_1 / c_0$$
$$= r_1.$$

This approximate estimator for $\hat{\alpha}_1$ is also intuitively appealing since r_1 is an estimator for $\rho(1)$ and $\rho(1) = \alpha_1$ for a first-order AR process. A confidence interval for α_1 may be obtained from the fact that the asymptotic standard error of $\hat{\alpha}_1$ is $\sqrt{\{(1 - \alpha_1^2)/N\}}$ although the confidence interval will not be symmetric for $\hat{\alpha}_1$ away from zero (Bartlett, 1966). In particular a test for $\alpha_1 = 0$ may be obtained, as suggested by the results of Section 4.1, by seeing if $\hat{\alpha}_1$ lies within the range $\pm 2/\sqrt{N}$.

For a second-order AR process, with $m = 2$, similar approximations may be made to give

$$\hat{\mu} \simeq \bar{x}$$
$$\hat{\alpha}_1 \simeq r_1 (1 - r_2)/(1 - r_1^2) \tag{4.10}$$
$$\hat{\alpha}_2 \simeq (r_2 - r_1^2)/(1 - r_1^2) \tag{4.11}$$

These results are also intuitively reasonable in that if we fit a second-order model to what is really a first order process, then as $\alpha_2 = 0$, we have $\rho(2) = \rho(1)^2 = \alpha_1^2$ and so $r_2 \simeq r_1^2$. Thus equations (4.10) and (4.11) become $\hat{\alpha}_1 \simeq r_1$ and $\hat{\alpha}_2 \simeq 0$. Jenkins and Watts (1968; p. 197) describe $\hat{\alpha}_2$ as the (sample) *partial* autocorrelation coefficient of order 2 which measures the excess correlation between $\{X_t\}$ and $\{X_{t+2}\}$ not accounted for by $\hat{\alpha}_1$.

In addition to point estimates of α_1 and α_2 it is also possible to find a confidence region in the (α_1, α_2) plane (Jenkins and Watts, 1968, p. 192).

Higher order AR processes may also be fitted by least squares in a straighforward way. Two alternative approximate methods are commonly used. Both methods involve taking

$$\hat{\mu} = \bar{x}.$$

The first method fits the data to the model

$$X_t - \bar{x} = \alpha_1 (x_{t-1} - \bar{x}) + \ldots + \alpha_m (x_{t-m} - \bar{x}) + Z_t$$

treating it as if it were an ordinary regression model. A standard multiple regression computer program may be used with appropriate modification.

The second method involves substituting the sample autocorrelation coefficients into the first m Yule–Walker equations (see Section 3.4.4) and solving for $(\hat{\alpha}_1, \ldots, \hat{\alpha}_m)$ (e.g. Pagano, 1972). In matrix form these equations are

$$\mathbf{R}\hat{\alpha} = \mathbf{r} \tag{4.12}$$

where

$$\mathbf{R} = \begin{pmatrix} 1 & r_1 & r_2 & \cdots & r_{m-1} \\ r_1 & 1 & r_1 & \cdots & r_{m-2} \\ r_2 & r_1 & 1 & \cdots & r_{m-3} \\ \hline r_{m-1} & r_{m-2} & \cdots & & 1 \end{pmatrix}$$

is an $(m \times m)$ matrix

$$\hat{\alpha}^T = (\hat{\alpha}_1, \ldots, \hat{\alpha}_m)$$

and

$$\mathbf{r}^T = (r_1, \ldots, r_m).$$

For N reasonably large, both methods will give estimated values 'very close' to the true least squares estimates for which $\hat{\mu}$ is close to but not necessarily equal to $\bar{x}$.

4.2.2 Determining the order of an autoregressive process

We may already have a good idea of the order of the process from the sample autocorrelation function. If not, a simple procedure for determining the order of an AR process is to fit processes of progressively higher order to the data, to calculate the residual sum of squares for each order, and to plot this against the order. For order zero, the residual sum of squares is the total corrected sum of squares $\Sigma(x_t - \bar{x})^2$. The residual sum of squares always decreases as the number of parameters is increased but it is often possible to see where the curve flattens out and the addition of extra parameters makes little improvement to the fit. As in polynomial regression, a visual inspection of this type may be supplemented by standard significance tests on the reduction of the residual sum of squares. Alternatively Jenkins and Watts (1968) suggest plotting the residual variance for processes of different order.

Another aid to determining the order of an AR process is the *partial autocorrelation function* (Jenkins and Watts, 1968, Section 5.4.3). When fitting a model of order m, the last coefficient, α_m, will be denoted by π_m and measures excess correlation at lag m not accounted for by a model of order $(m - 1)$. It is called the mth partial autocorrelation coefficient. If π_m is plotted against m, we have a graph of the sample partial autocorrelation function. Values which are significantly different from zero (outside the range $\pm 2/\sqrt{N}$) will indicate the order of the process, since the partial ac.f. of an mth-order AR process is zero for all lags greater than m. (However, note that the partial ac.f. of a first-order MA process decreases exponentially). An examination of the partial ac.f. will occasionally be helpful in identifying an appropriate model.

4.3 Fitting a moving average process

Suppose now that a MA process is thought to be an appropriate model for a given time series. As for an AR process, we have two problems:

(a) Finding the order of the process,

and

(b) Estimating the parameters of the process.

We consider problem (b) first.

4.3.1 Estimating the parameters of a moving average process

Estimation problems are more difficult for an MA process than an AR process, because efficient explicit estimators cannot be found. Instead some form of numerical iteration must be performed.

Let us begin by considering the first-order MA process

$$X_t = \mu + Z_t + \beta_1 Z_{t-1} \tag{4.13}$$

where μ, β_1 are constants and Z_t denotes a purely random process. We would like to write the residual sum of squares, ΣZ_t^2, solely in terms of the observed x's and the parameters μ, β_1, as we did for the AR process, differentiate with respect to μ and β_1 and hence find the least-squares estimates. Unfortunately the residual sum of squares is not a quadratic function of the parameters and so explicit least-squares estimates cannot be found. Nor can we simply equate sample and theoretical first-order autocorrelation coefficients by

$$r_1 = \hat{\beta}_1 / (1 + \hat{\beta}_1^2) \tag{4.14}$$

and choose the solution $\hat{\beta}_1$ such that $|\hat{\beta}_1| < 1$, because this gives rise to an inefficient estimator. (Whittle, 1953).

The approach suggested by Box and Jenkins (1970, Chapter 7) is as follows. Select suitable starting values for μ and β_1 such as $\mu = \bar{x}$ and β_1 given by the solution of equation 4.14. (See Table A in Box and Jenkins, 1970). Then the corresponding residual sum of squares may be calculated using (4.13) recursively in the form

$$Z_t = X_t - \mu - \beta_1 Z_{t-1} \qquad (4.15)$$

With $z_0 = 0$, we have $z_1 = x_1 - \mu$, $z_2 = x_2 - \mu - \beta_1 z_1, \ldots$,

$$z_N = x_N - \mu - \beta_1 z_{N-1}.$$

Then $\overset{N}{\underset{t=1}{\Sigma}} z_t^2$ may be calculated.

This procedure can then be repeated for other values of μ and β_1 and the sum of squares Σz_t^2 computed for a grid of points in the (μ, β_1) plane. We may then determine by inspection the least-squares estimates of μ and β_1 which minimize Σz_t^2. These least-squares estimates are also maximum likelihood estimates conditional on a fixed value of z_0 provided that Z_t is normally distributed. The procedure can be further refined by *back forecasting* the value of z_0 (see Box and Jenkins, 1970) but this is unnecessary except when β_1 is 'close' to plus or minus one.

An alternative estimation procedure due to J. Durbin is to fit a high order AR process to the data and use the duality between AR and MA processes (see for example Kendall, 1973, Section 12.8). This procedure has the advantage of requiring less computation, but the widespread availability of high-speed computers has resulted in the procedure becoming obsolete.

For higher-order processes a similar type of iterative procedure to that described above may be used. For example with a second-order MA process, one would guess starting values for μ, β_1, β_2, compute the residuals recursively using

$$z_t = x_t - \mu - \beta_1 z_{t-1} - \beta_2 z_{t-2}$$

and compute Σz_t^2. Then other values of μ, β_1, β_2 would be tried, probably over a grid of points, until the minimum value of Σz_t^2 is found. Clearly a computer is essential for performing such a large number of arithmetic operations, and for processes higher than second order some form of hill-climbing procedure may be desirable to minimize the residual sum of squares in a way that is numerically efficient. Box and Jenkins (1970, Section 7.2) describe such a procedure which they call 'non-linear estimation'. This description arises from the fact that the residuals are non-linear functions of the parameters but the description may give rise to confusion.

For a completely new set of data, it is usually a good idea to use the graphical method based on evaluating the residual sum of squares at a grid of points. A visual examination of the sum of squares surface will sometimes provide useful information. In particular it is interesting to see how 'flat' the surface is; if the surface is approximately quadratic; and if the parameter estimates are approximately uncorrelated.

In addition to point estimates, a confidence region for the MA parameters may also be found if we assume that the Z_t are normally distributed. Suppose our model contains p unknown parameters, θ. Let $\hat{\theta}$ denote the maximum likelihood estimate of θ and let $S(\hat{\theta})$, $S(\theta)$ denote the residual sum of squares at $\hat{\theta}$ and at θ. If $\text{Var}(Z_t) = \sigma_z^2$, then $S(\hat{\theta})/\sigma_z^2$ is distributed as χ^2 with $(N - p)$ degrees of freedom, while $[S(\theta) - |S(\hat{\theta})|]/\sigma_z^2$ is distributed as χ^2 with p degrees of freedom. Thus, if the model is correct,

$$\frac{[S(\theta) - (\hat{\theta})]p}{S(\hat{\theta})/N - p)}$$

is distributed as F with p and $(N - p)$ degrees of freedom. Thus the $100(1 - \alpha)\%$ confidence region for θ contains all those points θ in p-space such that

$$S(\theta) \leqslant S(\hat{\theta}) \left(1 + \frac{p}{N-p} F_{p,N-p,\alpha}\right)$$

If the residual sum of squares has been determined at a grid of points θ in p-space, then this confidence region may be easily obtained.

4.3.2 Determining the order of a moving average process

If a MA process is thought to be appropriate for a given set of data, the order of the process is usually evident from the sample ac.f. in view of the simple form of the theoretical ac.f. of a MA process (see Section 3.4.2). In principle one could compute the residual sum of squares or residual variance for different order MA processes and see where the graph levels out (Jenkins and Watts, 1968, Section 5.4.4) but this requires a lot of computing and is usually unnecessary.

4.4 Estimating the parameters of a mixed model

Suppose now that a mixed autoregressive-moving average (ARMA) model is thought to be appropriate for a given time series. The estimation problems for an ARMA model are similar to those for a MA model in that an iterative procedure has to be used. The residual sum of squares can be calculated at every point on a suitable grid of the parameter values, and the values which give the minimum sum of squares may then be assessed. Alternatively a numerical procedure such as hill-climbing may be used to arrive at the minimum sum of squares in a more efficient way. A confidence region may also be found as in the last section.

As an example consider an ARMA process of order $(1,1)$ given by

$$X_t - \mu = \alpha_1 (X_{t-1} - \mu) + Z_t + \beta_1 Z_{t-1}.$$

73

Given N observations $x_1, \ldots, x_N$, we guess values for μ, α_1, β_1, set $z_0 = 0$ and $x_0 = \mu$, and then calculate the residuals recursively by

$$z_1 = x_1 - \mu$$

$$z_2 = x_2 - \mu - \alpha_1 (x_1 - \mu) - \beta_1 z_1$$

$$\cdots \cdots \cdots$$

$$z_N = (x_N - \mu) - \alpha_1 (x_{N-1} - \mu) - \beta_1 z_{N-1}$$

The residual sum of squares $\sum\limits_{t=1}^{N} z_t^2$ may then be calculated. Then other values of μ, α_1, β_1 may be tried until the minimum residual sum of squares is found.

Further details may be found in Box and Jenkins (1970). Alternative approaches to this problem are given by Hannan (1970) and Akaike (1973a).

4.5 Estimating the parameters of an integrated model

In practice, most time series are non-stationary and the stationary models we have so far considered are not immediately appropriate. Then we can difference an observed time series until it is stationary as described in Section 3.4.6. An AR, MA or mixed model may then be fitted to the differenced series as described in Sections 4.2 – 4.4. The resulting model for the undifferenced series is called an integrated model.

4.6 The Box–Jenkins seasonal model

In practice, many time series contain a seasonal periodic component which repeats every s observations. For simplicity we consider monthly observations where $s = 12$, but the application to other values of s is straightforward. Box and Jenkins (1970) have generalized the ARIMA model to deal with seasonality and define a general multiplicative seasonal model in the form (using their notation)

$$\phi_p(B)\Phi_P(B^{12})w_t = \theta_q(B)\Theta_Q(B^{12})a_t \qquad (4.16)$$

where B denotes the backward shift operator, ϕ_p, Φ_P, θ_q, Θ_Q are polynomials of order p, P, q, Q respectively, and $\{a_t\}$ are independent random variables with mean zero and variance σ_a^2. The shift operator B^{12} is such that

$$B^{12}w_t = w_{t-12}.$$

For example if $p = Q = 1$, and $P = q = 0$, then (4.16) becomes

$$(1 - \alpha B)w_t = (1 + \beta B^{12})a_t$$

so that $w_t = \alpha w_{t-1} + a_t + \beta a_{t-12}$. The reader should compare the seasonal model (4.16) with equations (3.5a) and (3.6).

Equation 4.16 defines a stationary model provided that the roots of

$$\phi_p(B)\Phi_P(B^{12}) = 0$$

lie outside the unit circle. In order to fit the model to a non-stationary series, Box and Jenkins (1970) suggest differencing the original series to remove both trend and seasonality by

$$w_t = \nabla^d \nabla_{12}^D x_t$$

where

$$\nabla_{12}x_t = x_t - x_{t-12}$$

and

$$\nabla\nabla_{12}x_t = \nabla_{12}x_t - \nabla_{12}x_{t-1} = x_t - x_{t-1}$$
$$- x_{t-12} + x_{t-13}.$$

The values of the integers d and D do not usually need to exceed one.

Further remarks on fitting seasonal models may be found in Section 5.2.4, the subject being of particular interest for forecasting.

4.7 Residual analysis

When a model has been fitted to a time series, it is advisable to check that the model really does provide an adequate description of the data. The usual way of doing this is to examine the *residuals,* which are the differences between the observations and the fitted values. For example with a first order AR model (equation 3.4), the true errors are given by

$$z_t = x_t - \alpha x_{t-1}$$

if we know the true value of the parameter α. When α is estimated by least squares, the fitted value is $\hat{\alpha} x_{t-1}$, and the residual is given by

$$\hat{z}_t = x_t - \hat{\alpha} x_{t-1}.$$

The analysis of residuals is discussed by Box and Jenkins (1970; Section 8.2). It is important to realize that the residuals are necessarily correlated even if the true errors are independent. From Section 4.1, we have that the correlogram of the true errors is such that each r_k is approximately normally distributed, mean 0, variance $1/N$, for reasonably large values of N. But the correlogram of the residuals has somewhat different properties. For example for a first-order AR process with $\alpha = 0.7$, the 95% confidence limits are at $\pm 1.3/\sqrt{N}$ for r_1, $\pm 1.7/\sqrt{N}$ for r_2, and $\pm 2/\sqrt{N}$ for values of r_k at higher lags. Thus for lags greater than 2 the confidence limits *are* the same as for the correlogram of the true errors, but for lags 1 and 2 they are not.

This type of problem was first investigated by J. Durbin and G. S. Watson in regard to multiple regression models applied to time-series data (Durbin and Watson, 1950, 1951, 1971). Suppose we have $(k + 1)$ time series of N observations with a dependent variable y and k 'independent' variables $x_1, \ldots, x_k$ and we fit the model

$$y_t = \beta_1 x_{1t} + \ldots + \beta_k x_{kt} + Z_t \qquad (t = 1, \ldots, N).$$

Having estimated the parameters β_i by least squares we want to see if the errors Z_t are independent. The residuals

$$\hat{z}_t = y_t - \hat{\beta}_1 x_{1t} - \ldots - \hat{\beta}_k x_{kt} \qquad (t = 1, \ldots, N)$$

are therefore calculated and the Durbin–Watson statistic is given by

$$d = \frac{\sum\limits_{t=2}^{N} (\hat{z}_t - \hat{z}_{t-1})^2}{\sum\limits_{t=1}^{N} \hat{z}_t^2} \qquad (4.17)$$

The distribution of d under the null hypothesis that the Z_t are independently distributed has been investigated and tables are available (e.g. Kendall, 1973) which give critical values of d. These depend on the number of independent variables in the model.

It is useful to note that d is related to the first autocorrelation coefficient of the residuals. From (4.17) we have

$$\sum\limits_{t=2}^{N} (\hat{z}_t - \hat{z}_{t-1})^2 \simeq 2 \sum\limits_{t=1}^{N} \hat{z}_t^2 - 2 \sum\limits_{t=2}^{N} \hat{z}_t \hat{z}_{t-1}$$

so that

$$d \simeq 2(1 - r_1)$$

where $r_1 = \Sigma \hat{z}_t \hat{z}_{t-1} / \Sigma \hat{z}_t^2$ is the first autocorrelation coefficient of the residuals (since mean residual is zero).

A test on d is therefore asymptotically equivalent to a test on r_1. A statistic similar to d can also be calculated to test for higher-order autocorrelation. Wallis (1972) considers the statistic

$$d_4 = \sum_{t=5}^{N} (\hat{z}_t - \hat{z}_{t-4})^2 / \sum_{t=1}^{N} \hat{z}_t^2$$

for testing for fourth order autocorrelation. Here we find $d_4 \simeq 2(1 - r_4)$.

Durbin and Watson make it clear that their statistic should only be used when the independent variables can be regarded as 'fixed' variables. Unfortunately the test has sometimes been used inappropriately for autoregressive processes, and as Nerlove and Wallis (1966) have pointed out, a non-significant value of d does not preclude the possibility that the errors are positively autocorrelated if the model contains autoregressive terms (or lagged endogenous variables in econometric jargon). However a significant value of d *does* indicate autocorrelated errors both for multiple regression and AR models. Durbin (1970) has considered the Durbin–Watson test for models containing autoregressive terms, but a more general approach to the problem is provided by Box and Pierce (1970), who consider the large sample properties of all the residual autocorrelation coefficients for any ARMA process. Their results, usefully summarized by Box and Jenkins (1970), show that $1/\sqrt{N}$ supplies an upper bound for the standard error of the r_k's calculated from the residuals. For low lags, particularly lag 1, the standard error of r_k may be substantially less than $1/\sqrt{N}$ but for higher lags $1/\sqrt{N}$ is a good approximation. Thus, if we calculate the residual correlogram in the usual way, values of r_k which lie outside the range $\pm 2/\sqrt{N}$ are certainly significantly different from zero. But for low lags, values of r_k inside the range $\pm 2/\sqrt{N}$ may still be significantly different from zero and need further investigation.

Box and Jenkins (1970, Section 8.2.2) also describe what they call a portmanteau lack-of-fit test. Instead of looking at the residual r_k's separately, one looks at the first K values of

r_k together, by calculating

$$Q = N \sum_{k=1}^{K} r_k^2 \qquad (4.18)$$

where N is the number of terms in the differenced series. If the fitted model is appropriate, then Q should be approximately distributed as χ^2 with $(N - p - q)$ degrees of freedom, where p, q are the number of AR and MA terms, respectively. The test based on Q seems to have rather poor power properties. For example Chatfield and Prothero (1973a) fitted 4 different ARIMA models to the same set of data but all 4 models gave a non-significant value of Q.

Box and Jenkins (1970, Section 8.2.4) also suggest looking for periodicities in the residuals by examining a function of the data called the periodogram, which will be described in Chapter 7, and they describe a Kolmogorov–Smirnov test for the cumulative periodogram. However, Durbin (1973) has queried their statement that, for large samples, the periodogram for the estimated residuals will have similar properties to that for the true errors.

4.8 General remarks on model-building

How do we set about finding a suitable model for a given time series? The answer depends on a number of factors including the properties of the series as assessed by a visual examination of the data, the number of observations available, and the way the model is to be used.

In Chapter 3 we introduced a number of different types of stochastic process including ARMA and ARIMA processes. I agree with Box and Jenkins (1970) that this is a most useful general class of processes which provide a good fit to many different types of time series and which should generally be considered when more than 50 observations are available. Nevertheless other types of model are sometimes more useful, as for example in oceanography where trend and seasonal models are sometimes appropriate.

In areas such as oceanography and electrical engineering, long stationary series often occur. If a parametric model is required, an ARMA model should be considered and can be fitted as outlined earlier in the chapter. The observed correlogram and the partial ac.f. are examined, the appropriate ARMA model identified, and the model parameters estimated by least squares. However, as we shall see in Chapters 6 and 7, we may be more interested in the frequency properties of the time-series in which case an ARMA model may not be very helpful.

In many other areas, such as economics and marketing, non-stationary series often occur and in addition may be fairly short. If more than 50 observations are available, Box and Jenkins (1970) advocate the fitting of ARIMA models by differencing the observed time series until it becomes stationary and then fitting an ARMA model to the differenced series. For seasonal series, the seasonal ARIMA model may be used. However it should be clearly recognized that when the variation of the systematic part of the time series (i.e. the trend and seasonality) is dominant, the effectiveness of the ARIMA model is mainly determined by the initial differencing operations and not by the subsequent fitting of an ARMA model to the differenced series, even though the latter operation is much more time-consuming. Thus the simple models discussed in Chapter 2 may be perfectly adequate for some time series with a pronounced trend and/or large seasonal effect. For example we might represent a series by a linear trend together with seasonal and error terms in the form

$$x_t = a + bt + s_t + \epsilon_t$$

where a, b are constants. Models of this type have the advantage of simplicity, of being easy to interpret, and of being fairly robust. In addition they can be used for short series where it is impossible to fit an ARIMA model. But they are restrictive with regard to the form of the trend and usually assume the errors to be independent which is sometimes not the case.

Exercises

1. Derive the least-squares estimates for a first-order AR process having mean μ.

2. Derive the least-squares normal equations for an mth-order AR process, taking $\hat{\mu} = \bar{x}$, and compare with the Yule–Walker equations (equation 4.12).

3. For a second-order AR process, show that the (theoretical) partial autocorrelation coefficient of order 2 is given by

$$[\rho(2) - \rho(1)^2]/[1 - \rho(1)^2].$$

   Compare with equation (4.11).

4. Suppose that the correlogram of a time series consisting of 100 observations has $r_1 = 0.61$, $r_2 = 0.47$, $r_3 = -0.05$, $r_4 = 0.06$, $r_5 = -0.21$, $r_6 = 0.11$, $r_7 = 0.08$, $r_8 = 0.05$, $r_9 = 0.12$, $r_{10} = -0.01$. Suggest an ARMA model which may be appropriate.

CHAPTER 5

Forecasting

5.1 Introduction

Forecasting the future values of an observed time series is an important problem in many areas, including economics, production planning, sales forecasting and stock control.

Suppose we have an observed time series, $x_1, x_2, \ldots, x_N$. Then the problem is to estimate x_{N+1}, or more generally x_{N+q}. The prediction of x_{N+q} made at time N of the value q steps ahead will be denoted by $\hat{x}(N, q)$. The integer q is called the *lead time*. There is, of course, no universally applicable forecasting procedure. Instead we must look for the procedure which is most appropriate for a given set of conditions.

The many types of forecasting procedure can be classified into three broad categories:

(a) Subjective   Forecasts can be made on a subjective basis using judgement, intuition, commercial knowledge and any other relevant information. Procedures of this type, such as the Delphi method, will not be described here. (See, for example, Chambers *et al.*, 1971; Keay, 1972).

(b) Univariate   Forecasts can be based entirely on past observations in a given time series, by fitting a model to the data and extrapolating. For example, forecasts of future sales of a product would be based entirely on past sales. Methods

of this type are sometimes called 'naive' or *'projection'* methods.

(c) Multivariate   Forecasts can be made by taking observations on other variables into account. For example, sales may depend on stocks. Regression models are of this type, as are econometric models. The use of a 'leading indicator' also comes into this category. Multivariate models are sometimes called 'causal' or 'prediction' models.

In practice, a forecasting procedure may involve a *combination* of the above approaches. For example univariate forecasts are often computed, and then adjusted subjectively. Davis (1974) describes a marketing forecast as the result of bringing together the projections or predictions developed statistically from past data, and the knowledge and the art, of people deeply involved in the market.

Another type of combination is that proposed by Bates and Granger (1969) where two or more objective forecasts are computed and then a weighted average is calculated.

Before choosing a forecasting procedure, it is essential to consider how the forecast is to be used, what accuracy is required, how much money is available, how many items are to be forecast, how much data is available, and how far ahead forecasts are required. We will return to these questions in Section 5.4.

The forecasting procedure may depend somewhat on the required lead time. We will be mainly concerned with what is called *short-term* forecasting where the lead time is less than about nine months. For example in stock control, the lead time for which forecasts are required is the time between ordering an item and its delivery, which is usually a few weeks or months. Longer term forecasts are more often required in government and corporate planning and then it is generally advisable to use a fairly simple procedure. The longer the lead time, the more likely is the underlying model to change and hence the more unreliable the forecast.

Some forecasting procedures simply produce *point* forecasts. But these do not indicate the uncertainty associated with the estimation of future demand. Thus it is usually desirable to produce an *interval* forecast. Some procedures, such as Box—Jenkins, enable one to do this but others do not, in which case upper and lower limits may have to be assessed on a subjective basis.

5.2 Univariate procedures

This section introduces the many projection methods which are now available. Other useful references are Coutie *et al.* (1964), Brown (1963), and Kendall (1973).

For all the procedures, the first step is to *plot* the data as much useful information can often be obtained from a visual examination of the data, and this may help to suggest an appropriate forecasting procedure.

5.2.1 Extrapolation of trend curves

For long-term forecasting it is often useful to fit a trend curve to successive yearly totals and extrapolate. This problem is discussed by Gregg *et al.* (1964) and Harrison and Pearce (1972). A variety of curves may be tried including polynomial, exponential and Gompertz curves. At least seven to ten years of historical data are required and Harrison and Pearce suggest that 'one should not make forecasts for a longer period ahead than about half the number of past years for which data are available'.

For seasonal data, the trend and seasonal variation may be estimated by the methods of Chapter 2, either by using moving averages or by fitting a trend curve. The fitted curve may then be extrapolated into the future. The method is simple, fairly crude, but is robust and cheap and is worth considering for long-term forecasting where it is unlikely to be worthwhile fitting a complicated model to past data as the model may change in the future.

A drawback to the use of trend curves is that there is no

logical basis for choosing among the different curves except by goodness-of-fit. Unfortunately it is often the case that one can find several curves which fit a given set of data almost equally well, but which when projected forward, give widely different forecasts.

5.2.2 Exponential smoothing   This forecasting procedure, first suggested by C. C. Holt in about 1958, should only be used in its basic form for non-seasonal time series showing no trend. Of course many time series which arise in practice do contain a trend or seasonal pattern, but these effects can be measured and removed to produce a stationary series. Thus it turns out that adaptations of exponential smoothing can be used for most types of time-series and so the following remarks provide a useful basis for a study of the methods due to Brown (1963), Holt–Winters, Harrison (1965), and Box and Jenkins (1970).

Given a stationary, non-seasonal, time series, $x_1, x_2, \ldots, x_N$, it is natural to take as an estimate of x_{N+1} a weighted sum of the past observations.

$$\hat{x}(N, 1) = c_0 x_N + c_1 x_{N-1} + c_2 x_{N-2} + \ldots \qquad (5.1)$$

where the $\{c_i\}$ are weights. It seems sensible to give more weight to recent observations and less weight to observations further in the past. An intuitively appealing set of weights are geometric weights which decrease by a constant ratio. In order that the weights sum to one, we take

$$c_i = \alpha(1 - \alpha)^i \quad (i = 0, 1, \ldots)$$

where α is a constant such that $0 < \alpha < 1$. Then (5.1) becomes

$$\hat{x}(N, 1) = \alpha x_N + \alpha(1 - \alpha)x_{N-1} + \alpha(1 - \alpha)^2 x_{N-2} + \ldots. \qquad (5.2)$$

Strictly speaking, equation (5.2) implies an infinite number of past observations, but in practice there will only be a finite number. So equation (5.2) is customarily rewritten as

$$\hat{x}(N, 1) = \alpha x_N + (1 - \alpha)[\alpha x_{N-1} + \alpha(1 - \alpha)x_{N-2} + \ldots]$$
$$= \alpha x_N + (1 - \alpha)\hat{x}(N - 1, 1). \qquad (5.3)$$

If we set $\hat{x}(1, 1) = x_1$, then equation (5.3) can be used recursively to compute forecasts. Equation (5.3) also reduces the amount of arithmetic involved since forecasts can easily be updated using only the latest observation and the previous forecast.

The procedure defined by equation (5.3) is called exponential smoothing and has proved very useful in many forecasting situations. The adjective 'exponential' arises from the fact that the geometric weights lie on an exponential curve, but the procedure could equally well be called geometric smoothing.

Equation (5.3) is sometimes rewritten in the form

$$\hat{x}(N, 1) = \alpha[x_N - \hat{x}(N - 1, 1)] + \hat{x}(N - 1, 1)$$
$$= \alpha e_N + \hat{x}(N - 1, 1) \qquad (5.4)$$

where $\quad e_N = x_N - \hat{x}(N - 1, 1)$
$$= \text{prediction error at time } N.$$

It can be shown (Box and Jenkins, 1962) that exponential smoothing is optimum if the underlying model for the time series is given by

$$X_t = \mu + \alpha \sum_{j < t} Z_j + Z_t \qquad (5.5)$$

This infinite moving average process is non-stationary, but the first differences $(X_{t+1} - X_t)$ form a first-order moving average process.

The value of the smoothing constant, α, depends on the properties of the given time series. Values between 0.1 and 0.3 are commonly used and produce a forecast which depends on a large number of past observations. Values close to one are used rather less often and give forecasts which depend much more on recent observations. When $\alpha = 1$, the forecast is equal to the most recent observation.

The value of α may be estimated from past data by a similar procedure to that used for estimating the parameters of a moving average process. The sum of squared prediction errors is computed for different values of α and the value is chosen which minimizes the sum of squares. With a given value of α, calculate

$$\hat{x}(1, 1) = x_1$$

$$e_2 = x_2 - \hat{x}(1, 1)$$

$$\hat{x}(2, 1) = \alpha e_2 + \hat{x}(1, 1)$$

$$e_3 = x_3 - \hat{x}(2, 1)$$

$$\cdot \cdot \cdot \cdot \cdot \cdot \cdot \cdot \cdot \cdot \cdot \cdot \cdot \cdot$$

$$e_N = x_N - \hat{x}(N - 1, 1)$$

and compute $\sum\limits_{i=2}^{N} e_i^2$. Repeat this procedure for other values of α between 0 and 1, say in steps of 0.1, and select the value which minimizes Σe_i^2, (or omit the first few e's and minimize $\sum\limits_{i=c}^{N} e_i^2$). Usually the sum of squares surface is quite flat near the minimum and so the choice of α is not critical (Cox, 1961).

5.2.3 Holt—Winters forecasting procedure

Exponential smoothing may easily be generalized to deal with time series containing trend and seasonal variation. The resulting procedure is usually referred to as the Holt—Winters procedure and is described by Coutie *et al.* (1964) and Winters (1960). Trend and seasonal terms are introduced which are also updated by exponential smoothing.

Suppose the observations are monthly. Let m_t denote the estimated current mean in month t, r_t denote the estimated trend term in month t (i.e. the expected increase or decrease per month in the current mean), and s_t denote the estimated seasonal factor appropriate to month t. Then, as each new observation becomes available, all three terms are updated. If the seasonal variation is multiplicative (model B of Section 2.4), the updating equations are:

$$m_t = \alpha x_t / s_{t-12} + (1 - \alpha)(m_{t-1} + r_{t-1})$$
$$s_t = \beta x_t / m_t + (1 - \beta)s_{t-12}$$
$$r_t = \gamma(m_t - m_{t-1}) + (1 - \gamma)r_{t-1}.$$

where x_t is the latest observation and α, β, γ are constants such that $0 < \alpha, \beta, \gamma < 1$. The forecasts from time t are then

$$\hat{x}(t, h) = (m_t + hr_t)s_{t-12+h} \quad (h = 1, 2, \ldots, 12)$$

If the seasonal variation is additive, the updating equations are:

$$m_t = \alpha(x_t - s_{t-12}) + (1 - \alpha)(m_{t-1} + r_{t-1})$$
$$s_t = \beta(x_t - m_t) + (1 - \beta)s_{t-12}$$
$$r_t = \gamma(m_t - m_{t-1}) + (1 - \gamma)r_{t-1}.$$

A graph of the data should be examined to see if an additive or multiplicative seasonal effect is the more appropriate. The method may 'blow up' if the wrong model is used. If the seasonal period does not cover 12 observations, then the equations need to be modified in an obvious way. Starting values for m_t, r_t and s_t may be estimated in a fairly crude way from the first two years data, by taking

$$m_1 = \sum_{t=1}^{12} x_t/12, r_1 = (\text{mean of 2nd year} - \text{mean of 1st year})/12$$

and $s_1, \ldots, s_{12}$ to be the average seasonal effects in the first two years when the different months are compared with the yearly means. The three smoothing constants, α, β, and γ, are chosen as in Section 5.2.2 except that on this occasion the quantity to be minimized will be $\sum\limits_{i=25}^{N} e_i^2$.

If the seasonal smoothing constant, β, is close to zero, then the updating equations for s_t may need modification. The updating technique for exponential smoothing effectively assumes an infinite amount of past data. But if say six years monthly data is available, then each of the monthly factors will only be revised five times and inordinate weight may be given to the starting value. For example if $\beta = 0.1$ and the starting value for month 1 is s_1, then

$$s_{13} = 0.1x_{13}/m_{13} + 0.9s_1$$

and we eventually find

$$s_{61} = 0.1x_{61}/m_{61} + 0.09x_{49}/m_{49} + \ldots + 0.9^5 s_1.$$

This undesirable effect can be removed by using higher values for β in the first few years. One strategy is to take $\beta = 1/n$ in year n up to a maximum of about $n = 8$. Apart from the choice between multiplicative and additive seasonality, the Holt–Winters procedure can be made fully automatic in the sense that a single computer program can be written so as to produce forecasts for any given time series without human intervention, and the technique is widely used in industry.

5.2.4 Box–Jenkins forecasting procedure

This forecasting procedure has been developed by G. E. P. Box and G. M. Jenkins over the last 15 years. The reader is recommended to read Box and Jenkins (1968) before tackling their book (1970). Some other publications which describe the Box–Jenkins procedure and give examples are Naylor et al. (1972), Chatfield and Prothero (1973a), Thompson and Tiao (1971), and Nelson (1973). This section aims to give a broad outline of the procedure. Further details may be found in the above references.

The procedure consists basically of fitting a mixed autoregressive integrated moving average (ARIMA) model to a given set of data and then taking conditional expectations. The main stages in setting up a Box—Jenkins forecasting model are as follows:

(a) Model identification   Examine the data to see which member of the class of ARIMA processes appears to be most appropriate.

(b) Estimation   Estimate the parameters of the chosen model by least squares as described in Chapter 4.

(c) Diagnostic checking   Examine the residuals from the fitted model to see if it is adequate.

(d) Consider alternative models if necessary   If the first model appears to be inadequate for some reason, then other ARIMA models may be tried until a satisfactory model is found.

Now AR, MA and mixed models have been around for many years and are associated in particular with G. U. Yule and H. Wold. The major contribution of Box and Jenkins has been to provide a general *strategy* for time-series forecasting. By stressing differencing, they enable one to set up models for non-stationary series and the general class of ARIMA models is often referred to as the Box—Jenkins class of models. They have also provided general estimation and diagnostic checking procedures.

The first step in the Box—Jenkins procedure is to difference the data until it is stationary. This is achieved by examining the correlograms of various differenced series until one is found which comes down to zero 'fairly quickly' and from which any seasonal effect has been largely removed. For non-seasonal data, first-order differencing is usually sufficient. For seasonal data of period 12, the operator $\nabla \nabla_{12}$ is often used if the seasonal effect is additive, while the operator ∇_{12}^2 may be used if the seasonal effect is multiplicative. (See exercise 1.) Sometimes the operator ∇_{12} by itself will be sufficient. Over-differencing should be avoided. For quarterly data, the operator ∇_4 may be used, and so on.

The differenced series will be denoted by $\{w_t; t = 1, \ldots, N-c\}$ where c terms are 'lost' be differencing. For example, if the operator $\nabla\nabla_{12}$ is used, then $c = 13$.

If the data is non-seasonal, an ARMA model can now be fitted to $\{w_t\}$ as described in Chapter 4. If the data is seasonal, then the general seasonal ARMA model defined in equation 4.16 may be fitted as follows. 'Reasonable' values of p, P, q, Q are selected by examining the correlogram of the differenced series $\{w_t\}$. Values of p and q are selected as outlined in Chapter 4 by examining the first few values of r_k. Values of P and Q are selected by examining the values of r_k at $k = 12, 24 \ldots$ (where the seasonal period is 12). If for example r_{12} is 'large' but r_{24} is 'small', this suggests one seasonal moving average term so we would take $P = 0$, $Q = 1$.

Having tentatively identified what appears to be a reasonable seasonal ARMA model, least-squares estimates of the model parameters may be obtained by minimizing the residual sum of squares in a similar way to that proposed for ordinary ARMA models. In the case of seasonal series, it is advisable to estimate initial values of a_t and w_t by back-forecasting (or backcasting) rather than set them equal to zero. This procedure is described by Box and Jenkins (1970, Section 9.2.4). In fact if the model contains a seasonal moving average parameter which is close to one, several cycles of forward and backward iteration may be needed.

For both seasonal and non-seasonal data, the adequacy of the fitted model should be checked by what Box and Jenkins call 'diagnostic checking'. This essentially consists of examining the residuals from the fitted model to see if there is any evidence of non-randomness. The correlogram of the residuals is calculated and we can then see how many coefficients are significantly different from zero and whether any further terms are indicated for the ARIMA model.

If the fitted model appears to be inadequate, then alternative ARIMA models may be tried until a satisfactory model is found. When a satisfactory model is found, forecasts may readily be computed. For example suppose $p = 1$, $P = 0$, $q = 0$, $Q = 1$ so that

$$(1 - \alpha B)w_t = (1 + \beta B^{12})a_t$$

or

$$w_t = \alpha w_{t-1} + a_t + \beta a_{t-12}.$$

Given data up to time t, the best estimate of a_{t+1} is zero so that

$$\hat{w}(t, 1) = \hat{\alpha}w_t + \hat{\beta}\hat{a}_{t-11}$$

where $\hat{\alpha}$, $\hat{\beta}$ are estimates of α, β, and $\hat{a}_{t-11}$ is the estimated residual at time $(t-11)$.

This gives us a forecast of w_{t+1}, but of course we want a forecast of x_{t+1}. If for example

$$w_t = \nabla_{12}x_t = x_t - x_{t-12}$$

then we find

$$\hat{x}(t, 1) = x_{t-11} + \hat{\alpha}(x_t - x_{t-12}) + \hat{\beta}\hat{a}_{t-11}.$$

Forecasts for more than one step ahead may also be easily found.

In contrast to the Holt—Winters forecasting procedure which assumes a simple family of models, the Box—Jenkins procedure is not fully automatic. A model is chosen from the large class of ARIMA models according to the properties of the particular time-series being analysed. Thus while the procedure is more versatile than its competitors it is also more complicated and expensive, and considerable experience is required to identify an appropriate ARIMA model. One difficulty that sometimes arises is that several

ARIMA models may be found which fit the data equally well. Such models will often be almost equivalent and will therefore yield nearly identical forecasts, but this is not always the case. Another drawback to the method is that at least 50, and preferably 100, observations are needed for it to have a chance of success. We will make further comments on the applicability of the method in Section 5.4.

5.2.5 Stepwise autoregression The main disadvantage of the Box–Jenkins approach is that it is a fairly complicated procedure which requires considerable skill on the part of the statistician as well as some lengthy computer programs. Newbold and Granger (1974) have described a procedure called stepwise autoregression which is in some sense a subset of the Box–Jenkins procedure and has the advantage of being fully automatic and of utilizing a standard multiple regression computer program.

First differences of the data are taken to allow for non-stationarity in the mean. Then a maximum possible lag, say p, is chosen. The best autoregressive model with just one coefficient is then found:—

$$w_t = \mu + \alpha_s^{(1)} w_{t-s} + e_t^{(1)} \qquad 1 \leqslant s \leqslant p.$$

where $w_t = x_t - x_{t-1}$, and $\alpha_s^{(1)}$ is the autoregression coefficient. Then the best autoregressive model with 2 coefficients, 3 coefficients, etc., is found. The procedure is terminated when the reduction in the sum of squared residuals at the jth stage is less than some pre-assigned quantity.

Thus an integrated autoregressive model is fitted which is a special case of the Box–Jenkins ARIMA class.

Newbold and Granger suggest choosing $p = 13$ for quarterly data and $p = 25$ for monthly data.

5.2.6 Other methods Several other forecasting procedures have been proposed. Brown (1963) has suggested a technique called *general exponential smoothing* which consists of fitting polynomial, sinusoidal or exponential functions to the data and finding appropriate updating formula. One special case of this is double exponential smoothing which is applicable to series containing a linear trend. Note that Brown suggests fitting by *discounted* least squares, in which more weight is given to recent observations. Generally speaking, Brown's method compares unfavourably with other univariate forecasting procedures, especially on seasonal data (see Reid, 1972).

P. J. Harrison has suggested two modifications to seasonal exponential smoothing. The first (1965) consists essentially of performing a Fourier analysis of the seasonal factors and replacing them by smoothed factors. The second (Harrison and Stevens, 1971) is a Bayesian approach which uses different smoothing constants according to whether the series is showing 'normal' behaviour or if there is evidence of a step change in the mean level of the series or in the trend. When there is evidence of a change, the smoothing constants are chosen closer to one so as to allow the mean and trend terms to adapt quicker. Prior probabilities of the system being normal or in a state of change have to be specified. Such a procedure is worth considering for series which tend to show step changes in mean or trend. The method is related to a technique used in control theory called Kalman filtering (Astrom, 1970).

A forecasting method proposed by Bloomfield (1973) depends on approximating the logarithm of an estimated spectral density function (see Chapter 7) by a truncated Fourier series. First results indicate that the method tends to perform rather less well than the Box—Jenkins method. Another technique called adaptive filtering has been proposed by Wheelwright and Makridakis (1973) but this technique appears to be technically unsound (Chatfield and Newbold, 1974).

5.3 Multivariate procedures

This section provides a brief introduction to multivariate forecasting procedures.

5.3.1 Multiple regression This approach uses the multiple linear regression model where the variable of interest (say y) is linearly related to present and past values of other variables (say $x_1, \ldots, x_p$) and also possibly to past values of y (these are autoregressive terms).

A description of multiple regression can be found in many statistics textbooks (e.g. Anderson, 1971) and will not be repeated here. With the general availability of multiple regression computer programs it is computationally easy (perhaps too easy!) to fit a multiple regression model and use it for planning or forecasting.

Multiple regression models sometimes work well, particularly in a marketing context. But there are several dangers in the method which need to be appreciated. Firstly the ready availability of computer programs has resulted in a tendency to put more and more explanatory variables into the model with dubious results. The resulting model may indeed appear to give a good fit to the available data. By including say 20 explanatory variables one may achieve a multiple correlation coefficient, R^2, as high as 0.995, but this good fit may be spurious and does not necessarily mean that the model will give good forecasts. A more sensible number of explanatory variables is a maximum of 6 or 7, and it is advisable to fit the model to part of the available data and then check the model by forecasting the remainder of the data.

In marketing applications the so-called 'independent' variables are usually not independent at all, and if some of them are highly correlated there may be singularity problems. It is therefore advisable to look at the correlation matrix of the 'independent' variables *before* carrying out a multiple regression so that, if necessary, some variables can be excluded. It is unnecessary for the explanatory variables to be completely independent, but large correlations should be avoided.

Another difficulty arising in multiple regression is that some crucial explanatory variables may have been held more or less constant in the past and it is then impossible to assess their effect and include them in the model in a quantitative way. For example a company may be considering increasing its advertising expenditure and would like to construct a model which would predict the effect on sales. But if advertising has been held relatively constant in the past, then it will be impossible to estimate the effect of advertising, and yet a model which excludes advertising may be useless if advertising expenditure is changed.

But perhaps the most important problem in multiple regression forecasting concerns the structure of the error terms. It is often assumed that these are an independent white noise sequence, but such an assumption is sometimes not appropriate and may lead to disappointing results (Granger and Newbold, 1974), or to an inappropriate model (Box and Newbold, 1971). Having fitted a multiple regression model, one should check the residuals for autocorrelation as described in Section 4.7. If the residuals are autocorrelated, one can try fitting a multiple regression model with autocorrelated errors by a method, due to D. Cochrane and G. H. Orcutt, which is described by Kendall (1973; Section 12.9). Alternatively one can try the Box—Jenkins approach.

In summary, I am inclined to agree with Brown (1963, p. 77) that the use of multiple regression models can be very dangerous except in certain special cases where one has a definite reason why one series should be related to another. An example of such a situation, involving just one explanatory variable, is given in example 3 of Section 5.5. Another example is given by Bhattacharyya (1974).

5.3.2 Econometric models   Econometric models (e.g. Christ, 1966) assume that an economic system can be described, not by a single equation, but by a set of simultaneous equations. For example, not only do wage rates depend on prices but also prices depend on wage rates. Economists distinguish between exogenous variables, which affect the system but are not themselves affected, and endogenous variables which interact with each other. The simultaneous equations may be written in the form

$$Ax_t + By_t + \sum_j B_j y_{t-j} = u_t$$

where

x_t = an $m \times 1$ vector of exogenous variables

y_t = an $n \times 1$ vector of endogenous variables

u_t = an $n \times 1$ vector of random error terms

$A, B, \{B_j\}$ = matrices of parameters which have to be estimated by econometric methods. Each matrix has n rows so that the number of equations equals the number of endogenous variables.

It is a large scale operation to set up an econometric model and we shall say no more about it here. A simple example, involving just two equations, is given by Leuthold *et al.* (1970).

5.3.3 Box–Jenkins method   Box and Jenkins (1968 and 1970) consider multivariate forecasting as well as univariate forecasting. They propose a class of models, called transfer function models (see Chapter 9), which is a natural extension of the ARIMA class. They concentrate on the case of describing the relationship between one 'output' variable (e.g. sales) and one 'input' variable (e.g. advertising expenditure), though the method can be generalized to cover several input variables. They show how to identify an appropriate model and then make predictions. This method will be introduced in Section 9.4.2.

5.4 A comparison of forecasting procedures

Many factors need to be considered when choosing the forecasting procedure which is most appropriate for a given set of conditions. These factors include the purpose of the forecast, the degree of accuracy required, and the amount of money available. No clear-cut rules can be given but the following remarks are intended as a general guide.

The most important consideration is how the forecast is to be used. Many forecasts are used for planning purposes, for example in production planning or stock control. Alternatively, a forecast may act as a 'norm' or yardstick. Sometimes more than one forecast may be required. For example suppose that a company wants to carry out an advertising campaign so as (hopefully) to increase sales. Then they may require a 'norm' forecast based only on past sales so that any increase in sales may be assessed, and may also require a planning forecast based on all available information which will try to forecast the effect of the sales campaign so as to ensure that ample product is available.

Other considerations in choosing a forecasting procedure include the required accuracy of the forecasts, how much account should be taken of changes in other factors, and how important the past is thought to be in determining the future. If the model is thought likely to change, then it is not worth spending too much time fitting a complicated model to past data. Whatever forecasting method is adopted, one is still extrapolating from the past into the future, with all the inherent dangers. Clearly, one should always be prepared to modify forecasts in the light of any other information.

As they do not take other variables into account, univariate forecasts are mainly intended to act as a 'norm'. If forecasts are required for planning or decision-making, then ideally one should set up a multivariate model. However, multivariate models are relatively difficult and expensive to build, particularly econometric and transfer function models. It is also fair to say that econometric models have often turned out to have poor predictive powers. For example, Naylor *et al.* (1972) show that univariate Box–Jenkins forecasts are better than forecasts obtained using the Wharton econometric model. Thus although one would expect multivariate models to give better forecasts than univariate models, this is not necessarily so.

Of the univariate procedures, recent work by D. J. Reid, C. W. J. Granger and P. Newbold suggests that the Box–Jenkins method tends to give more accurate forecasts than other methods. However, the Box–Jenkins method is also more expensive than other methods and so the cost of forecasting must be balanced against accuracy. Holt–Winters, Harrison and step-wise autoregression are roughly equal in accuracy but better than Brown's method. Newbold and Granger (1974) found that a combination of Holt–Winters and step-wise autoregression (which are both automatic procedures) gave results which were almost as accurate as those given by the Box–Jenkins approach.

The choice of forecasting procedure also depends on the number of items which have to be forecast. If a large number of items are involved, say several hundred, then for production planning and stock control purposes it may be useful to set up a computer forecasting system to allow routine forecasts to be made without human intervention. Then a univariate procedure is appropriate. The high initial cost of the Box–Jenkins procedure will make it unsuitable in this sort of situation. Instead, one will probably prefer a fully automatic procedure such as Holt—Winters or step-wise autoregression.

For a company that manufactures only a few items, Coutie *et al.* (1964) suggest that the salesman concerned can probably produce the best forecasts by subjective methods. This is sometimes frowned upon in academic circles as being 'unscientific', but there is some evidence to suggest that subjective methods may often be superior though the evidence is not all one way. Gearing (1971) gets better results using a rule-of-thumb technique than with an adaptive smoothing technique. Wagle *et al.* (1968) found several products for which hand forecasts are markedly better than objective statistical forecasts. Armstrong and Grohman (1972) discuss an example where econometric forecasts are better than univariate forecasts, which in turn are better than subjective forecasts. Myers (1971) says that forecasting should never be entirely objective and recommends extrapolating by eye and then estimating the effects of external factors on a subjective basis. My own preference in sales forecasting would generally be to compute univariate forecasts using a simple procedure such as Holt—Winters and then, if desired, to adjust the forecasts subjectively.

Finally let us attempt an assessment of the Box—Jenkins procedure, which is of great interest in the statistical world at the present time. Most univariate forecasting procedures are fully automatic in that only one set of forecasts can be produced from a given set of data. However, the Box—Jenkins procedure involves a subjective element which allows one to choose from a wide class of models. This greater versatility is both the strength and weakness of the Box—Jenkins approach. The advantage of being able to choose from a wide class of models rather than being restricted to one particular model is clear. At the same time, the subjective assessment involved in choosing a model means that considerable experience is required in interpreting sample autocorrelation functions. In addition the procedure is expensive in terms of both computing and staff time.

In many time-series situations, such as analysing macro-economic series, a large expenditure of time and effort can certainly be justified and then the method is well worth considering, provided that sufficient expertise is available. However for routine sales forecasting we take the view that the method is usually inappropriate for several reasons. Firstly, when variation is dominated by trend and seasonality, the effectiveness of the ARIMA model is mainly determined by the differencing procedure and some other way of measuring trend and seasonality may be more appropriate. Secondly, many firms do not have sufficient expertise available to employ the method effectively: Thirdly, the method may be too expensive: Fourthly, a fully automatic procedure is often appropriate: Fifthly, a manager is more likely to accept statistical forecasts if he understands the way they have been derived. Most managers understand the ideas of trend and seasonality, but would be completely baffled by ARIMA models. As G. J. A. Stern has pointed out, to introduce Box–Jenkins methods in some firms would require translating managers from the statistical stone age to the statistical space age in one leap, which is unrealistic. Nevertheless the ideas behind the Box–Jenkins method are undoubtedly of interest to any statistician interested in forecasting. With suitable development, they should become a useful tool in the time-series analyst's kit. For example Bhattacharyya (1974) gives an example combining regression and Box–Jenkins.

5.5 Some examples

In this section we discuss 3 sets of data, to illustrate some of the problems which arise in real forecasting situations.

Example 1 Fig. 5.1 shows the (coded) sales of a certain company in successive quarters over 6 years. Suppose that a

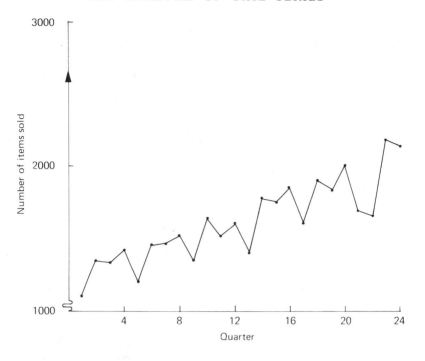

Figure 5.1   Sales of a certain company in successive three-month periods.

univariate forecast is required for the next four quarters. What method is appropriate? This series demonstrates the importance of plotting a time series and making a visual examination before deciding on the appropriate forecasting procedure. It is evident from Fig. 5.1 that there is an increasing trend and a pronounced seasonal effect with observations 1, 5, 9, 13, . . ., relatively low. The series is too short to use the Box–Jenkins method. Instead a suitable forecasting procedure might appear to be that of Holt–Winters. But closer examination of Fig. 5.1 reveals that the observation in quarter 22 is unusually low while the following observation seems somewhat high. If we were to apply Holt–Winters with no modification, these unusual observations would have a marked effect on the forecasts. We must therefore first decide if they indicate a permanent change in the seasonal pattern, in which case earlier observations will have little relevance for forecasting purposes, or if they were caused by some unusual phenomenon such as a strike. In the latter case, some adjustment of the observations in quarters 22 and 23 needs to be made before forecasts are computed.

Example 2   Fig. 5.2 shows some telephone data analysed by Tomasek (1972) using the Box–Jenkins method. He developed the model

$$(1 - 0.84B)(1 - B^{12})(x_t - 132)$$
$$= (1 - 0.60B)(1 + 0.37B^{12})a_t,$$

which when fitted to all the data explained 99.4% of the total variation about the mean. [i.e. the total corrected sum of squares, $\Sigma(x_t - \bar{x})^2$]. On the basis of this good fit, Tomasek recommended the use of the Box–Jenkins method for forecasting.

However, it is not at all clear that this is a sensible recommendation. Looking at Fig. 5.2, we see that the series has an unusually high regular seasonal pattern. In fact 97% of the variation about the mean is explained by a linear trend and constant seasonal pattern. As we remarked in Section 4.8, when the variation due to trend and seasonality is dominant, the effectiveness of the ARIMA model is mainly determined by the initial differencing operations and not by the time-consuming ARMA model fitting to the differenced series (Akaike, 1973b). For such regular data, nearly any forecasting method will give good results. For example the Holt–Winters method explains 98.9% of the variation and it is rather doubtful if the extra expense of the Box–Jenkins method can be justified by increasing the explained variation from 98.9% to 99.4%.

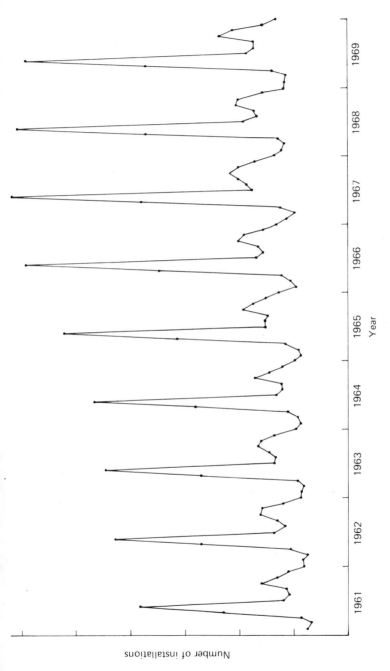

Figure 5.2   Tomasek's telephone data. Numbers of newly installed telephones in successive months. (Tomasek does not provide a vertical scale.)

104

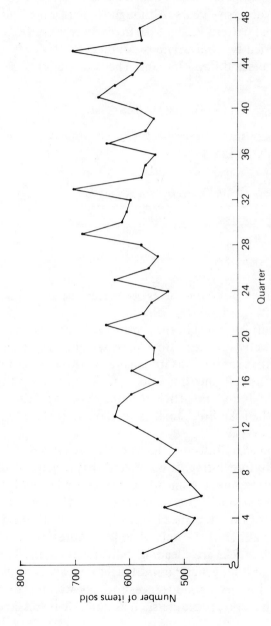

Figure 5.3   Sales of company C in successive three-month periods.

Example 3 Fig. 5.3 shows quarterly sales data for company C over 12 successive years. Although there is some evidence of a seasonal pattern it is not particularly regular. D. L. Prothero tried two univariate procedures on this data, namely Holt—Winters and Box—Jenkins. The Box—Jenkins model fitted was

$$\nabla \nabla_4 x_t = (1 - 0.2B)(1 - 0.8B^4)a_t$$

The mean absolute forecast errors up to four quarters ahead were calculated as follows.

No. of quarters ahead	Holt—Winters	Box—Jenkins
1	2543	2454
2	2864	2988
3	3174	3416
4	3732	4143

Thus although Box—Jenkins is 3% better one step ahead, it is up to 10% worse four steps ahead.

This result is typical in the sense that Box—Jenkins tends to do less well as the lead time increases. But the result is not typical in that for series as irregular as that shown in Fig. 5.3, Box—Jenkins will sometimes do considerably better than other methods and, if sufficient money and expertise are available, then the Box—Jenkins method is worth a try for data of this type.

This data also illustrates the possible advantages of multivariate forecasting. It was found that if detrended, deseasonalized sales are linearly regressed on detrended, deseasonalized stocks two quarters before, the mean absolute forecast error one step ahead was 1900 which is considerably better than either of the univariate procedures we tried. In other words stocks are a leading indicator for sales. This illustrates another general point that if one wants to put in a lot of effort to get a good forecast, it may well be better to try a multivariate procedure such as multiple regression rather than a complicated univariate procedure such as Box–Jenkins, although this does not always follow (Naylor *et al.*, 1972).

5.6 Prediction theory

Over the last thirty years or so, a general theory of linear prediction has been developed by Kolmogorov, Wiener (1949), Grenander and Rosenblatt (1957), Yaglom (1962), Whittle (1963) and Bartlett (1966) among others. All these authors avoid the use of the word 'forecasting' although most of the univariate methods considered in Section 5.2 are in the general class of linear predictors. The theory of linear prediction has applications in control and communications engineering and is of considerable theoretical interest, but readers who wish to tackle the sort of forecasting problem we have been considering earlier in this chapter will find this literature less accessible than the other references. Here we will only give a brief introduction.

Two types of problem are often distinguished. In the first type of problem we have data up to time T, $\{x_T, x_{T-1} \ldots\}$, and wish to predict the value of x_{T+m}. One approach is to use the predictor

$$\hat{x}_{T+m} = \Sigma c_j x_{T-j}$$

which is a linear function of the available data. The weights $\{c_j\}$ are chosen so as to minimize the expected mean square prediction error, $E(x_{T+m} - \hat{x}_{T+m})^2$. This is often called the *prediction* problem (e.g. Cox and Miller, 1968), while Yaglom (1962) refers to it as the *extrapolation* problem, and Whittle (1963) calls it *pure prediction*. As an example of the sort of result which has been obtained, Wiener (1949) has considered the problem of evaluating the weights $\{c_j\}$, so as to find the best linear predictor, when the ac.f. of the series $\{x_t\}$ is *known* and when the entire past of the sequence $\{x_t\}$ is known. It is interesting to compare this sort of approach with the forecasting techniques proposed earlier in this chapter. The Box–Jenkins approach, for example, also employs a linear predictor which will be optimal for a particular ARIMA process. But whereas Wiener says little about estimation, Box and Jenkins show how to find a linear predictor when the ac.f. has to be estimated.

The second type of problem arises when the process of interest, $s(t)$, called the *signal*, is contaminated by *noise*, $n(t)$, and we actually observe the process

$$y(t) = s(t) + n(t).$$

In some situations the noise is simply measurement error. In engineering applications the noise is an interference process of some kind. The problem now is to separate the signal from the noise. Given measurements on $y(t)$ up to time T we may want to reconstruct the signal up to time T or alternatively make a prediction of $s(T + \tau)$. The problem of reconstructing the signal is often called *smoothing* or *filtering*. The problem of predicting the signal is also often called *filtering* (Yaglom, 1962; Cox and Miller, 1968), but is sometimes called *prediction* (Astrom, 1970; Laning and Battin, 1956). It is often assumed that the signal and noise processes are uncorrelated and that $s(t)$ and $n(t)$ have known ac.f's.

It is clear that both the above types of problem are closely related to the control problem because, if we can predict how a process will behave, then we can adjust the process so that the achieved values are, in some sense, as close as possible to the target value. Further remarks on control theory must await a study of linear systems.

Exercises

1. (a) If $x_t = (a + bt)s_t + \epsilon_t$ where a, b are constants, $s_t = s_{t-12}$, and ϵ_t is a white noise process, show that $\nabla_{12}^2 x_t$ is a stationary process.
   (b) If $x_t = a + bt + s_t + \epsilon_t$, show that $\nabla\nabla_{12} x_t$ is a stationary process.

2. For the Box–Jenkins seasonal ARIMA model with $p = 0$, $P = 1$, $q = 1$, $Q = 0$, and $w_t = \nabla_{12} x_t$, find forecasts at time N for up to 12 steps ahead in terms of observations and estimated residuals up to time N.

CHAPTER 6

Stationary Processes in the Frequency Domain

6.1 Introduction

In Chapter 3 we described several types of stationary stochastic process, placing emphasis on the autocovariance (or autocorrelation) function which is the natural tool for considering the evolution of a process through time. In this chapter we introduce a complementary function called the *spectral density function,* which is the natural tool for considering the frequency properties of a time series. Inference regarding the spectral density function is called an analysis in the *frequency domain.*

Some statisticians initially have difficulty in understanding the frequency approach, but the advantages of frequency methods are widely appreciated in such fields as electrical engineering, geophysics and meteorology. These advantages will become apparent in the next few chapters.

We shall confine ourselves to real-valued processes. Many authors consider the more general problem of complex-valued processes, and this results in some gain of mathematical conciseness. But, in my view, the reader is more likely to understand an approach restricted to real-valued processes. The vast majority of practical problems are covered by this approach.

6.2 The spectral distribution function

In order to introduce the idea of a spectral density function, we must first consider a function called the *spectral distribution function*. The approach adopted is heuristic and not mathematically rigorous, but will, hopefully, give the reader a better understanding of the subject than the more theoretical approach adopted for example by Grenander and Rosenblatt (1957).

Suppose we suspect that a time series contains a periodic component at a known frequency. Then a natural model is

$$X_t = R \cos(\omega t + \theta) + Z_t \tag{6.1}$$

where ω is called the *frequency* of the periodic variation, R is called the *amplitude* of the variation, θ is called the *phase*, and Z_t denotes some stationary random series. An example of the periodic component is shown in Fig. 6.1. Note that the angle $(\omega t + \theta)$ is measured in radians. Since

$$\omega = \text{number of radians per unit time}$$

it is sometimes called the *angular* frequency, but in keeping with most authors we will simply call ω the frequency. However some authors, notably Jenkins and Watts (1968), refer to frequency as

$$f = \omega/(2\pi)$$
$$= \text{number of cycles per unit time}$$

and this form of frequency is much easier to interpret from a physical point of view. We will usually use the angular frequency, ω, in mathematical formulae for conciseness, but will often use the frequency $f = \omega/(2\pi)$ for the interpretation of data. The period of a sinusoidal cycle, called the *wavelength*, is clearly $2\pi/\omega$ or $1/f$.

Model 6.1 is a very simple model but in practice the variation in a time series may be caused by variation at several different frequencies. For example, sales figures may contain weekly, monthly, yearly and other cyclical variation.

111

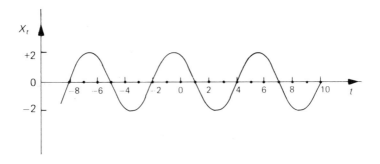

Figure 6.1   A graph of R cos($\omega t + \theta$) *with R = 2, ω = $\pi/3$, and*
$\theta = \pi/6$.

In other words the data show variation at high, medium and low frequencies. It is natural therefore to generalize (6.1) to:

$$X_t = \sum_{j=1}^{k} R_j \cos(\omega_j t + \theta_j) + Z_t \qquad (6.2)$$

where R_j is the amplitude at frequency ω_j.

The reader will notice that models (6.1) and (6.2) are *not* stationary if R, θ, $\{R_j\}$, and $\{\theta_j\}$ are fixed constants because $E(X_t)$ will change with time. In order to apply the theory of stationary processes to models like (6.1) and (6.2), it is customary to assume that R and $\{R_j\}$ are (uncorrelated) random variables with mean zero, or that θ and $\{\theta_j\}$ are random variables with a uniform distribution on $(0, 2\pi)$, which are fixed for a single realisation of the process (see Section 3.5 and exercise 14 of Chapter 3). This is something of a 'mathematical trick', but it does enable us to treat time series containing one or more deterministic sinusoidal components as stationary series.

Since cos($\omega t + \theta$) = cos ωt cos θ − sin ωt sin θ, model (6.2) is often expressed as a sum of sine and cosine terms in

112

the form

$$X_t = \sum_{j=1}^{k} (a_j \cos \omega_j t + b_j \sin \omega_j t) + Z_t \qquad (6.3)$$

where $a_j = R_j \cos \theta_j$

and $b_j = -R_j \sin \theta_j$.

But we may now ask why there should only be a finite number of frequencies involved in model (6.2) or (6.3). In fact, letting $k \to \infty$, the work of Wiener and others has shown that any discrete stationary process measured at unit intervals may be represented in the form

$$X_t = \int_0^{\pi} \cos \omega t \, du(\omega) + \int_0^{\pi} \sin \omega t \, dv(\omega) \qquad (6.4)$$

where $u(\omega)$, $v(\omega)$ are uncorrelated continuous processes with orthogonal increments (see Section 3.4.8) which are defined for all ω in the range $(0, \pi)$. Equation (6.4) is called the *spectral representation* of the process. The derivation of the spectral representation will not be considered here (see for example Cox and Miller, 1968, Chapter 8).

The reader may wonder why the upper limits of the integrals in (6.4) are π rather than ∞. For a continuous process, the upper limits would indeed by ∞, but for a discrete process measured at unit intervals of time there is no loss of generality in restricting ω to the range $(0, \pi)$ since

$$\cos(\omega t + k\pi t) = \begin{cases} \cos \omega t & k, \, t \text{ integers with } k \text{ even} \\ \cos(\pi - \omega)t & k, \, t \text{ integers with } k \text{ odd} \end{cases}$$

and so variation at frequencies higher than π cannot be distinguished from variation at the corresponding frequency in $(0,\pi)$. The frequency $\omega = \pi$, is called the *Nyquist frequency*. We will say more about this in Section 7.2.1. For a discrete process measured at equal intervals of time of length Δt, the Nyquist frequency is $\pi/\Delta t$. In the next two sections we consider discrete processes measured at unit intervals of time, but the arguments carry over to discrete processes measured at intervals Δt if we replace π by $\pi/\Delta t$.

The main point of introducing the spectral representation (6.4) is to show that every frequency in the range $(0, \pi)$ may contribute to the variation of the process. However the processes $u(\omega)$ and $v(\omega)$ in (6.4) are of little direct practical interest. Instead we shall introduce a function, $F(\omega)$, called the (power) spectral distribution function, which is related to $u(\omega)$ and $v(\omega)$, and which arises from a theorem (e.g. Bartlett, 1966, Section 6.1), called the Wiener–Khintchine theorem, named after N. Wiener and A. Y. Khintchine. As applied to real-valued processes, this theorem says that for any stationary stochastic process with autocovariance function $\gamma(k)$, there exists a monotonically increasing function, $F(\omega)$, such that

$$\gamma(k) = \int_0^\pi \cos \omega k \, dF(\omega). \qquad (6.5)$$

Equation (6.5) is called the spectral representation of the autocovariance function. It can be shown that the function, $F(\omega)$, has a direct physical interpretation, in that

$F(\omega)$ = contribution to the variance of the series which is accounted for by frequencies in the range $(0, \omega)$

It is most important to understand this physical interpretation of $F(\omega)$. There is no variation at negative frequencies so that

$F(\omega) = 0$ for $\omega < 0$

For a discrete process measured at unit intervals of time, the highest possible frequency is π and so all the variation is accounted for by frequencies less than π. Thus

$F(\pi) = \text{Variance}(X_t) = \sigma_X^2.$

This last result also comes directly from (6.5) with $k = 0$ when

$$\gamma(0) = \sigma_X^2 = \int_0^\pi dF(\omega).$$

In between $\omega = 0$ and $\omega = \pi$, $F(\omega)$ is monotonically increasing. In fact the function behaves rather like the cumulative distribution function of a probability distribution except that the upper limit is σ_X^2 rather than one.

If the process contains a deterministic sinusoidal component at frequency ω_0, say $R \cos(\omega_0 t + \theta)$ where R is a constant and θ is uniformly distributed on $(0, 2\pi)$, then there will be a step increase in $F(\omega)$ at ω_0 equal to $E[R^2 \cos^2(\omega_0 t + \theta)] = \tfrac{1}{2}R^2$.

As $F(\omega)$ is monotonic, it can be decomposed into two functions, $F_1(\omega)$ and $F_2(\omega)$ such that

$$F(\omega) = F_1(\omega) + F_2(\omega) \tag{6.6}$$

where $F_1(\omega)$ is a non-decreasing continuous function and $F_2(\omega)$ is a non-decreasing step function. This decomposition usually corresponds to the Wold decomposition with $F_1(\omega)$ relating to the purely indeterministic component of the process, while $F_2(\omega)$ relates to the deterministic component.

We shall be mainly concerned with purely indeterministic processes, where $F_2(\omega) \equiv 0$, so that $F(\omega)$ is a continuous function on $(0, \pi)$.

The adjective 'power', which is sometimes prefixed to 'spectral distribution function' derives from the engineer's use of the word in connection with the passage of an electric current through a resistance. For a sinusoidal input, the power is directly proportional to the squared amplitude of the oscillation. For a more general input, the power spectral distribution function describes how the power is distributed with respect to frequency. In the case of a time series, the variance may be regarded as the total power.

Note that some authors use a normalized form of $F(\omega)$

given by

$$F^*(\omega) = F(\omega)/\sigma_X^2 \qquad (6.7)$$
$$= \textit{proportion} \text{ of variance accounted for by}$$
$$\text{frequencies in the range } (0, \omega).$$

Since $F^*(\pi) = 1$, and $F^*(\omega)$ is monotonically increasing, $F^*(\omega)$ has similar properties to a cumulative distribution function.

6.3 The spectral density function

For a purely indeterministic discrete stationary process, the spectral distribution function is a continuous (monotone bounded) function in $(0, \pi)$, and may therefore be differentiated† with respect to ω in $(0, \pi)$. We will denote the derivative by $f(\omega)$ where

$$f(\omega) = \frac{\mathrm{d}F(\omega)}{\mathrm{d}\omega} \qquad (6.8)$$
$$= \text{(power) spectral density function.}$$

The term 'spectral density function' is often shortened to *spectrum.* The adjective 'power' is sometimes omitted.

When $f(\omega)$ exists, equation (6.5) can be expressed in the form

$$\gamma(k) = \int_0^\pi \cos \omega k \, f(\omega) \mathrm{d}\omega \qquad (6.9)$$

Putting $k = 0$, we then have

$$\gamma(0) = \sigma_X^2 = \int_0^\pi f(\omega) \mathrm{d}\omega = F(\pi). \qquad (6.10)$$

The physical meaning of the spectrum is that $f(\omega) \, \mathrm{d}\omega$ represents the contribution to variance of components with

---

†(Strictly speaking, $F(\omega)$ may not be differentiable on a set of measure zero, but this is of no practical importance.)

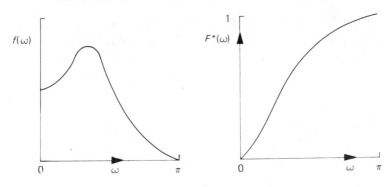

Figure 6.2   An example of a spectrum together with the corresponding normalized spectral distribution function.

frequencies in the range $(\omega, \omega + d\omega)$. When the spectrum is drawn, equation (6.10) indicates that the total area underneath the curve is equal to the variance of the process. A peak in the spectrum indicates an important contribution to variance at frequencies in the appropriate region. An example of a spectrum is shown in Fig. 6.2 together with the corresponding normalized spectral distribution function.

It is important to realize that the autocovariance function (acv.f) and the power spectral density function are equivalent ways of describing a stationary stochastic process. From a practical point of view, they are complementary to each other. Both functions contain the same information but express it in different ways. In some situations a time-domain approach based on the acv.f. is more useful while in other situations a frequency-domain approach is preferable. The relationship between the two functions is somewhat similar to that between the probability density function and the characteristic function of a continuous probability distribution.

Equation (6.9) expresses $\gamma(k)$ in terms of $f(\omega)$ as a cosine transform. The inverse relationship (see Appendix I) is given by

$$f(\omega) = \frac{1}{\pi} \sum_{k=-\infty}^{\infty} \gamma(k)e^{-i\omega k} \qquad (6.11)$$

117

so that the spectrum is the *Fourier transform* of the autocovariance function. Since $\gamma(k)$ is an even function, (6.11) is often written in the equivalent form

$$f(\omega) = \frac{1}{\pi}\left[\gamma(0) + 2\sum_{k=1}^{\infty}\gamma(k)\cos\omega k\right]. \tag{6.12}$$

Note that if we try to apply (6.12) to a process containing a deterministic component at frequency ω_0, then $\Sigma\gamma(k)\cos\omega_0 k$ will not converge, since $F(\omega)$ is not differentiable at ω_0 and so $f(\omega_0)$ is not defined.

The reader should be made aware at this point that several other definitions of the spectrum are given in the literature, most of which differ from (6.12) by a constant multiple and by the range of definition of $f(\omega)$. The most popular approach is to define the spectrum in the range $(-\pi, \pi)$ by

$$f(\omega) = \frac{1}{2\pi}\sum_{k=-\infty}^{\infty}\gamma(k)e^{-i\omega k} \tag{6.13}$$

whose inverse relationship (see Appendix I) is

$$\gamma(k) = \int_{-\pi}^{\pi}e^{i\omega k}f(\omega)\,d\omega \tag{6.14}$$

Jenkins and Watts (1968) use these equations except that they take $f = \omega/2\pi$ as the variable (see equations A3 and A4 in Appendix I). Equations (6.13) and (6.14), which form a Fourier transform pair, are the more usual form of the Wiener–Khintchine relations. The formulation is slightly more general in that it can be applied to complex valued time series. But for real time series, we find that $f(\omega)$ is an even function and then one need only consider $f(\omega)$ for $\omega > 0$. In my experience, the introduction of negative frequencies, while having certain mathematical advantages, serves only to confuse the student. As we are concerned only with real-valued processes, we prefer (6.11) defined on $(0, \pi)$.

It is sometimes useful to use a normalized form of the

spectral density function given by

$$f^*(\omega) = f(\omega)/\sigma_X^2$$
$$= \frac{\mathrm{d}F^*(\omega)}{\mathrm{d}\omega} \tag{6.15}$$

This is the derivative of the normalized spectral distribution function (see equation 6.7). Then we find that $f^*(\omega)$ is the Fourier transform of the *autocorrelation* function, namely

$$f^*(\omega) = \frac{1}{\pi} \left[1 + 2 \sum_{k=1}^{\infty} \rho(k) \cos \omega k \right], \tag{6.16}$$

and that $f^*(\omega)\,\mathrm{d}\omega$ is the *proportion* of variance between $(\omega, \omega + \mathrm{d}\omega)$. Kendall and Stuart (1966) and Kendall (1973) define the spectral density function in the range $(0, \pi)$ in terms of the autocorrelation function but omit the constant $1/\pi$ from equation (6.16). This makes it more difficult to give the function a physical interpretation. [Note that the inverse relationship (6.56) given by Kendall (1973) is incorrect.]

6.4 The spectrum of a continuous process

For a continuous purely indeterministic stationary process, $X(t)$, the autocovariance function $\gamma(\tau)$ is defined for all τ and the (power) spectral density function, $f(\omega)$, is defined for all positive ω. The relationship between these functions is very similar to that in the discrete case except that there is no upper bound to the frequency. We have

$$f(\omega) = \frac{1}{\pi} \int_{-\infty}^{\infty} \gamma(\tau)e^{-i\omega\tau} \, \mathrm{d}\tau$$

$$= \frac{2}{\pi} \int_{0}^{\infty} \gamma(\tau) \cos \omega\tau \, \mathrm{d}\tau \tag{6.17}$$

for $0 < \omega < \infty$, with the inverse relationship

$$\gamma(\tau) = \int_0^\infty f(\omega) \cos \omega\tau \, d\omega. \tag{6.18}$$

The integral in (6.17) converges for all ω provided that $X(t)$ is a purely indeterministic process.

6.5 Examples

In this section we derive the spectral density functions of some simple but important stationary processes.

(a) Purely random process A purely random process in discrete time, $\{Z_t\}$, is defined in Section 3.4.1. If $\mathrm{Var}(Z_t) = \sigma_Z^2$, then the acv.f. is given by

$$\gamma(k) = \begin{cases} \sigma_Z^2 & k = 0 \\ 0 & \text{otherwise} \end{cases}$$

so that the power spectral density function is given by

$$f(\omega) = \sigma_Z^2/\pi \tag{6.19}$$

using (6.12). In other words the spectrum is constant in the range $(0, \pi)$.

We have already pointed out that a continuous white noise process is physically unrealizable. A process is regarded as a practical approximation to continuous white noise if its spectrum is substantially constant over the frequency band of interest, even if it then approaches zero at high frequency.

(b) First-order moving average process The first-order MA process (see Section 3.4.3)

$$X_t = Z_t + \beta Z_{t-1}$$

has ác.f.

$$\rho(k) = \begin{cases} 1 & k = 0 \\ \beta/(1 + \beta^2) & k = \pm 1 \\ 0 & \text{otherwise.} \end{cases}$$

120

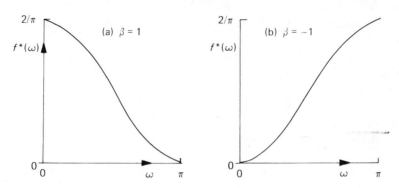

Figure 6.3   *Two examples of spectra of first-order moving average processes with (a) $\beta = 1$; (b) $\beta = -1$.*

So, using (6.16), the normalized spectral density function is given by

$$f^*(\omega) = \frac{1}{\pi} [1 + 2\beta \cos \omega/(1 + \beta^2)] \qquad (6.20)$$

for $0 < \omega < \pi$. The power spectral density function is then

$$f(\omega) = \sigma_X^2 f^*(\omega)$$

where $\sigma_X^2 = (1 + \beta^2)\sigma_Z^2$.

The shape of the spectrum depends on the value of β. When $\beta > 0$, the power is concentrated at low frequencies giving what is called a low-frequency spectrum, while if $\beta < 0$, power is concentrated at high frequencies giving a high frequency spectrum. Examples are shown in Fig. 6.3.

(c) First-order autoregressive process   The first-order AR process (see Section 3.4.4)

$$X_t = \alpha X_{t-1} + Z_t \qquad (6.21)$$

has acv.f.

$$\gamma(k) = \sigma_X^2 \alpha^{|k|} \qquad (k = 0, \pm 1, \pm 2, \ldots).$$

121

The power spectral density function is then, using (6.11),

$$f(\omega) = \frac{\sigma_X^2}{\pi} \left(1 + \sum_{k=1}^{\infty} \alpha^k e^{-ik\omega} + \sum_{k=1}^{\infty} \alpha^k e^{ik\omega} \right)$$

$$= \frac{\sigma_X^2}{\pi} \left(1 + \frac{\alpha e^{-i\omega}}{1 - \alpha e^{-i\omega}} + \frac{\alpha e^{i\omega}}{1 - \alpha e^{i\omega}} \right)$$

which after some algebra gives

$$f(\omega) = \sigma_X^2 (1 - \alpha^2)/[\pi(1 - 2\alpha \cos \omega + \alpha^2)] \qquad (6.22)$$

$$= \sigma_Z^2/[\pi(1 - 2\alpha \cos \omega + \alpha^2)] \qquad (6.23)$$

since $\sigma_Z^2 = \sigma_X^2(1 - \alpha^2)$.

The shape of the spectrum depends on the value of α. When $\alpha > 0$, power is concentrated at low frequencies while if $\alpha < 0$ power is concentrated at high frequencies. Examples are shown in Fig. 6.4.

It is hoped that the reader finds the shapes of the spectra in Fig. 6.4 intuitively reasonable. For example if α is negative then it is clear from (6.21) that values of X_t will tend to oscillate: a positive value of X_t will tend to be followed by a negative value and vice versa. These rapid oscillations correspond to high frequency variation.

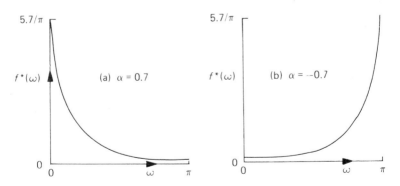

Figure 6.4   *Two examples of spectra of first-order autoregressive processes with (a) $\alpha = 0.7$; (b) $\alpha = -0.7$.*

(d) Higher-order autoregressive processes   It can be shown (e.g. Jenkins and Watts, 1968) that the spectrum of a second-order AR process with parameters α_1, α_2 is given by

$$f(\omega) = \sigma_Z^2 / \pi [1 + \alpha_1^2 + \alpha_2^2 - 2\alpha_1 (1 - \alpha_2) \cos \omega - 2\alpha_2 \cos 2\omega]$$

for $0 < \omega < \pi$. The shape of the spectrum depends on the values of α_1 and α_2. It is possible to get a high frequency spectrum, a low frequency spectrum, a spectrum with a peak between 0 and π, or a spectrum with a minimum between 0 and π.

For higher-order AR processes, one can get spectra with several peaks or troughs.

(e) A deterministic sinusoidal perturbation   Suppose that

$$X_t = \cos(\omega_0 t + \theta) \tag{6.24}$$

where ω_0 is a constant in $(0, \pi)$ and θ is a random variable which is uniformly distributed on $(0, 2\pi)$. As explained in Section 3.5, θ is fixed for a single realization of the process and (6.24) defines a purely deterministic process.

The acv.f. of the process is given by

$$\gamma(k) = \tfrac{1}{2} \cos \omega_0 k$$

which we note does *not* tend to zero as k increases. This is a feature of most deterministic processes.

From (6.24), it is obvious that all the 'power' of the process is concentrated at the frequency ω_0. Now $\mathrm{Var}(X_t) = E(X_t^2) = \tfrac{1}{2}$, so that the power spectral distribution function is given by

$$F(\omega) = \begin{cases} 0 & \omega < \omega_0 \\ \tfrac{1}{2} & \omega \geqslant \omega_0 \end{cases}$$

Since this is a step function, it has no derivative at ω_0 and so the spectrum is not defined at ω_0. If we nevertheless try to use equation (6.12) to obtain the spectrum as the Fourier transform of the acv.f. then we find that

$$f(\omega) = 0 \qquad \omega \neq \omega_0$$

but that $\Sigma \gamma(k) \cos \omega k$ does not converge at $\omega = \omega_0$. Jenkins and Watts (1968, p. 224) suggest writing the spectrum in terms of the Dirac delta function (see Appendix II), but there are mathematical difficulties involved in this approach.

(f) A mixture Our final example contains a mixture of deterministic and stochastic components, namely

$$X_t = \cos(\omega_0 t + \theta) + Z_t$$

where ω_0, θ, are as defined in example (e) above, and $\{Z_t\}$ is a purely random process, mean zero and variance σ_Z^2. Then we find that the acv.f. is given by

$$\gamma(k) = \begin{cases} \tfrac{1}{2} + \sigma_Z^2 & k = 0 \\ \tfrac{1}{2} \cos \omega_0 k & k = \pm 1, \pm 2, \ldots \end{cases}.$$

Again note that $\gamma(k)$ does *not* tend to zero as k increases because X_t contains a deterministic component.

We can obtain the power spectral distribution function by using (6.6), as the deterministic component, $\cos(\omega_0 t + \theta)$, has distribution function

$$F_1(\omega) = \begin{cases} 0 & \omega < \omega_0 \\ \tfrac{1}{2} & \omega \geq \omega_0 \end{cases}$$

while the stochastic component, Z_t, has a distribution function

$$F_2(\omega) = \sigma_Z^2 \omega / \pi \qquad (0 < \omega < \pi)$$

on integrating (6.19). Thus the overall spectral distribution function is given by

$$F(\omega) = \begin{cases} \sigma_Z^2 \omega / \pi & 0 < \omega < \omega_0 \\ \tfrac{1}{2} + \sigma_Z^2 \omega / \pi & \omega_0 \leq \omega < \pi \end{cases}.$$

As in example (e), the power spectrum is not defined at $\omega = \omega_0$.

Exercises

In the following questions, $\{Z_t\}$ denotes a purely random process, mean zero and variance σ_Z^2.

1. Find (a) the power spectral density function, (b) the normalized spectral density function of the first-order AR process:

$$X_t = \lambda X_{t-1} + Z_t.$$

2. Find the power spectral density functions of the following MA processes:

(a) $X_t = Z_t + Z_{t-1} + Z_{t-2}$

(b) $X_t = Z_t + 0.5Z_{t-1} - 0.3Z_{t-2}$

3. Show that the second-order MA process

$$X_t = \mu + Z_t + 0.8Z_{t-1} + 0.5Z_{t-2}$$

is second-order stationary, where μ denotes a constant. Find the acv.f. and ac.f. of $\{X_t\}$ and show that its normalized spectral density function is given by

$$f^*(\omega) = (1 + 1.27 \cos \omega + 0.53 \cos 2\omega)/\pi.$$
$$(0 < \omega < \pi)$$

4. A stationary time series $(X_t; t = \ldots, -1, 0, +1, \ldots)$ has normalized spectral density function

$$f^*(\omega) = 2(\pi - \omega)/\pi^2. \qquad (0 < \omega < \pi)$$

Show that its ac.f. is given by

$$\rho(k) = \begin{cases} 1 & k = 0 \\ (2/\pi k)^2 & k \text{ is odd} \\ 0 & k \text{ is even} (\neq 0) \end{cases}$$

125

5. A two-state Markov process may be set up as follows. Alpha particles from a radioactive source are used to trigger a flip-flop device which takes the states +1 and −1 alternately. The times t_i at which changes occur constitute a Poisson process, with mean event rate λ. Let $X(t)$ denote the state variable at time t. If the process is started at $t = 0$ with $P(X(0) = 1) = P(X(0) = -1) = \frac{1}{2}$, show that the process is second-order stationary, with autocorrelation function

$$\rho(u) = e^{-2\lambda|u|} \qquad (-\infty < u < \infty)$$

and spectral density function

$$f(\omega) = 4\lambda/[\pi(4\lambda^2 + \omega^2)] \qquad (0 < \omega < \infty).$$

CHAPTER 7

Spectral Analysis

Spectral analysis, sometimes called 'spectrum analysis', is the name given to methods of *estimating* the spectral density function, or spectrum, of a given time series.

In the last century, research workers, such as A. Schuster, were essentially concerned with looking for 'hidden periodicities' in the data, but spectral analysis as we know it today is mainly concerned with estimating the spectrum over the whole range of frequencies. M. S. Bartlett and J. W. Tukey have been prominent in the development of modern spectral analysis which has a history of about thirty years, and the techniques are now widely used by many scientists, particularly in electrical engineering, physics, meteorology and marine science.

We are mainly concerned with purely indeterministic processes, which have a continuous spectrum, but the techniques can also be used for deterministic processes to pick out periodic components in the presence of noise.

7.1 Fourier analysis

Spectral analysis is essentially a modification of *Fourier* analysis so as to make it suitable for stochastic rather than deterministic functions of time. Fourier analysis (e.g. Hsu, 1967) is basically concerned with approximating a function by a sum of sine and cosine terms, called the Fourier series representation. Suppose that a function $f(t)$ is defined on $(-\pi, \pi]$ and satisfies the Dirichlet conditions [i.e. that it is absolutely integrable over this range, has a finite number of discontinuities, and a finite number of maxima and minima]. Then $f(t)$ may be approximated by the Fourier series

$$\frac{a_0}{2} + \sum_{r=1}^{k} (a_r \cos rt + b_r \sin rt),$$

where $a_0 = \dfrac{1}{\pi} \displaystyle\int_{-\pi}^{\pi} f(t)\mathrm{d}t$

$\qquad a_r = \dfrac{1}{\pi} \displaystyle\int_{-\pi}^{\pi} f(t) \cos rt \ \mathrm{d}t \qquad (r = 1, 2, \dots)$

$\qquad b_r = \dfrac{1}{\pi} \displaystyle\int_{-\pi}^{\pi} f(t) \sin rt \ \mathrm{d}t. \qquad (r = 1, 2, \dots).$

It can be shown that this Fourier series converges to $f(t)$ as $k \to \infty$ except at points of discontinuity where it converges to $\frac{1}{2}[f(t + 0) + f(t - 0)]$.

In order to apply Fourier analysis to discrete time series, we need to consider the Fourier series representation of $f(t)$ when $f(t)$ is defined only on the integers $1, 2, \dots, N$. Rather than write down the formula, we shall see that the required Fourier series emerges naturally by considering a simple sinusoidal model.

7.2 A simple sinusoidal model

Suppose we suspect that a given time series, with observations made at unit time intervals, contains a deterministic sinusoidal component at frequency ω together with a random error term. So we will consider the model

$$X_t = \mu + \alpha \cos \omega t + \beta \sin \omega t + Z_t \qquad (7.1)$$

where Z_t denotes a purely random process, and μ, α, β are parameters to be estimated from the data.

The observations will be denoted by $(x_1, x_2, \dots, x_N)$. The algebra in the next few sections is somewhat simplified if we confine ourselves to the case where N is *even*. Some other authors (e.g. Jenkins and Watts, 1968) also impose this condition. There is no real difficulty in extending the results to the case where N is odd (e.g. Anderson, 1971), and indeed many of the later estimation formulae apply for both odd and even N, but some results require one to consider odd N and even N separately. For example it is a little disconcerting to find that the Fourier series representation of a discrete time series with N odd (e.g. Pocock, 1974) does not contain a component at the Nyquist frequency. Thus, if N happens to be odd and a spectral analysis is required, our preference is to remove the first observation so as to make N even. If N is reasonably large, little information is lost and there is some gain in simplicity.

The model (7.1) can be represented in matrix notation by

$$E(\mathbf{X}) = A\boldsymbol{\theta}$$

where

$$\mathbf{X}^T = (X_1, \ldots, X_N)$$
$$\boldsymbol{\theta}^T = (\mu, \alpha, \beta)$$

$$\text{and} \quad A = \begin{pmatrix} 1 & \cos \omega & \sin \omega \\ 1 & \cos 2\omega & \sin 2\omega \\ \cdots & \cdots & \cdots \\ 1 & \cos N\omega & \sin N\omega \end{pmatrix}$$

It is 'well-known' that the least-squares estimate of $\boldsymbol{\theta}$ which minimizes $\sum_{t=1}^{N} (x_t - \mu - \alpha \cos \omega t - \beta \sin \omega t)^2$ is given by (e.g. Jenkins and Watts, 1968, Appendix A4.1)

$$\hat{\boldsymbol{\theta}} = (A^T A)^{-1} A^T \mathbf{x}$$

where

$$\mathbf{x}^T = (x_1, \ldots, x_N).$$

Now the highest frequency we can fit to the data is the Nyquist frequency, given by $\omega = \pi$ (see Section 7.2.1), while the lowest frequency we can reasonably fit completes 1 cycle in the whole length of the time series (see Section 7.2.1). By equating the cycle length $2\pi/\omega$ to N, we find that this lowest frequency is given by $2\pi/N$. The least-squares estimates are particularly simple if ω is restricted to one of the values

$$\omega_p = 2\pi p/N \qquad (p = 1, \ldots, N/2)$$

as $(A^T A)$ then turns out to be a diagonal matrix in view of the following 'well-known' trigonometric results (all summations are for $t = 1$ to N):

$$\Sigma \cos \omega_p t = \Sigma \sin \omega_p t = 0 \qquad (7.2)$$

$$\Sigma \cos \omega_p t \cos \omega_q t = \begin{cases} 0 & p \neq q \\ N & p = q = N/2 \\ N/2 & p = q \neq N/2 \end{cases} \qquad (7.3)$$

$$\Sigma \sin \omega_p t \sin \omega_q t = \begin{cases} 0 & p \neq q \\ 0 & p = q = N/2 \\ N/2 & p = q \neq N/2 \end{cases} \qquad (7.4)$$

$$\Sigma \cos \omega_p t \sin \omega_q t = 0 \qquad \text{for all } p, q. \qquad (7.5)$$

With $(A^T A)$ diagonal, we can easily find $\hat{\boldsymbol{\theta}}$, and for ω_p ($p \neq N/2$) we find (exercise 2)

$$\left. \begin{aligned} \hat{\mu} &= \Sigma x_t/N = \bar{x} \\ \hat{\alpha} &= 2\Sigma x_t \cos(\omega_p t)/N \\ \hat{\beta} &= 2\Sigma x_t \sin(\omega_p t)/N \end{aligned} \right\} \qquad (7.6)$$

while if $p = N/2$ we ignore the term in $\beta \sin \omega t$ which is zero for all t and find

$$\left. \begin{aligned} \hat{\mu} &= \bar{x} \\ \hat{\alpha} &= \Sigma x_t(-1)^t/N \end{aligned} \right\} . \qquad (7.7)$$

The model (7.1) is essentially the one used in the last century to search for hidden periodicities, but this model has now gone out of fashion. However it can still be useful if we have reason to suspect that a time series does contain a deterministic periodic component at a known frequency and we wish to isolate this component (e.g. Granger and Hatanaka, 1964, p. 6). Pocock (1974) gives an example where this type of procedure is used to investigate the seasonal variation in sickness absence.

Readers who are familiar with the analysis of variance technique will see that the total corrected sum of squared deviations, namely

$$\sum_{t=1}^{N} (x_t - \bar{x})^2,$$

can be partitioned into two components which are the residual sum of squares and the sum of squares 'explained' by the periodic component at frequency ω_p. This latter component is given by

$$\sum_{t=1}^{N} (\hat{\alpha} \cos \omega_p t + \hat{\beta} \sin \omega_p t)^2$$

which after some algebra (exercise 2) can be shown to be

$$\begin{array}{ll} (\hat{\alpha}^2 + \hat{\beta}^2)N/2 & p \neq N/2 \\ \hat{\alpha}^2 N & p = N/2 \end{array} \qquad (7.8)$$

using (7.2)–(7.5).

7.2.1 The Nyquist frequency

In Chapter 6 we pointed out that for a discrete process measured at unit intervals there is no loss of generality in restricting the spectral distribution function to the range $(0, \pi)$. We now demonstrate that the upper bound, π, called the Nyquist frequency, is indeed the highest frequency about which we can get meaningful information from a set of data.

First we will give a more general form for the Nyquist frequency. If observations are taken at equal intervals of length Δt, then the Nyquist (angular) frequency is $\omega_N = \pi/\Delta t$. The equivalent frequency expressed in cycles per unit time is $f_N = \omega_N/2\pi = 1/2\Delta t$.

Consider the following example. Suppose that temperature readings are taken every day in a certain town at noon. It is clear that these observations will tell us nothing about temperature variation *within* a day. In particular they will not tell us if nights are hotter or cooler than days. With only one observation per day, the Nyquist frequency is $\omega_N = \pi$ radians per day or $f_N = 1$ cycle per two days. This is lower than the frequencies which account for variation within a day. For example variation with a wavelength of one day has (angular) frequency $\omega = 2\pi$ radians per day or $f =$ one cycle per day. In order to get information about variation within a day we must take two or more observations per day.

A similar example is provided by yearly sales figures which will obviously give no information about any seasonal effects, whereas monthly or quarterly observations *will* give information about seasonality.

Finally a comment about the lowest frequency we can fit to a set of data. If we had just six months of temperature readings from winter to summer it would not be clear if there was an upward trend in the observations or if winters are colder than summers. However with one year's data it *would* become clear that winters are colder than summers. Thus if we are interested in variations within a year at the comparatively low frequency of 1 cycle per year, then we must have at least one year's data. Thus the lower the frequency we are interested in, the longer the time period over which we need to take measurements, whereas the higher the frequency we are interested in, the more frequently must we take observations.

7.3 Periodogram analysis

Early attempts at discovering hidden periodicities in a given
time series basically consisted of repeating the analysis of
Section 7.2 at all the frequencies $2\pi/N$, $4\pi/N \ldots , \pi$. In view
of (7.3)–(7.5) the different terms are orthogonal and we end
up with the finite Fourier series representation of the $\{x_t\}$,
namely

$$x_t = a_0 + \sum_{p=1}^{N/2-1} [a_p \cos(2\pi pt/N) + b_p \sin(2\pi pt/N)]$$
$$+ a_{N/2} \cos \pi t, \qquad (t = 1, 2, \ldots , N) \tag{7.9}$$

where the coefficients $\{a_p, b_p\}$ are of the same form as
equations (7.6) and (7.7), namely

$$\left.\begin{array}{l} a_0 = \bar{x} \\ a_{N/2} = \Sigma(-1)^t x_t/N \\ \left.\begin{array}{l} a_p = 2[\Sigma\, x_t \cos(2\pi pt/N)]/N \\ b_p = 2[\Sigma\, x_t \sin(2\pi pt/N)]/N \end{array}\right\} \; p = 1, \ldots , N/2-1 \end{array}\right\} \tag{7.10}$$

An analysis along these lines is sometimes called a Fourier
analysis or a *harmonic* analysis. The Fourier series
representation (7.9) has N parameters to describe N
observations and so can be made to fit the data exactly (just
as a polynomial of degree $(N - 1)$ involving N parameters can
be found which goes exactly through N observations in
polynomial regression). This explains why there is no error
term in (7.9) in contrast to (7.1). Also note that there is no
term in $\sin \pi t$ in (7.9) as $\sin \pi t$ is zero for all integer t.

It is worth stressing that the Fourier series coefficients
(7.10) at a given frequency ω are exactly the same as the
least-squares estimates for model (7.1).

The overall effect of the Fourier analysis of the data is to
partition the variability of the series into components at
frequencies $2\pi/N$, $4\pi/N, \ldots , \pi$. The component at frequency

$\omega_p = 2\pi p/N$ is often called the pth harmonic. For $p \neq N/2$, it is often useful to write the pth harmonic in the equivalent form

$$a_p \cos \omega_p t + b_p \sin \omega_p t = R_p \cos(\omega_p t + \phi_p) \qquad (7.11)$$

where

R_p = *amplitude* of pth harmonic

$$= \sqrt{(a_p^2 + b_p^2)} \qquad (7.12)$$

and ϕ_p = *phase* of pth harmonic

$$= \tan^{-1}(-b_p/a_p). \qquad (7.13)$$

We have already noted in Section 7.2 that, for $p \neq N/2$, the contribution of the pth harmonic to the total sum of squares is given by $N(a_p^2 + b_p^2)/2$. Using (7.12), this is equal to $NR_p^2/2$. Extending this result using (7.2)–(7.5) and (7.9), we have, after some algebra, that (exercise 3)

$$\sum_{t=1}^{N} (x_t - \bar{x})^2 = N \sum_{p=1}^{N/2-1} R_p^2/2 + Na_{N/2}^2. \qquad (7.14)$$

Dividing through by N we have

$$\Sigma(x_t - \bar{x})^2/N = \sum_{p=1}^{N/2-1} R_p^2/2 + a_{N/2}^2, \qquad (7.15)$$

which is often known as Parseval's theorem. The left-hand side of (7.15) is effectively the variance of the observations although the divisor is N rather than the more usual $(N-1)$. Thus $R_p^2/2$ is the contribution of the pth harmonic to the variance, and (7.15) shows how the total variance is partitioned.

If we plot $R_p^2/2$ against $\omega_p = 2\pi p/N$ we obtain a line spectrum. A different type of line spectrum occurs in the physical sciences when light from molecules in a gas discharge tube is viewed through a spectroscope. The light has energy at discrete frequencies and this energy can be seen as bright lines. But most time series have continuous spectra and then it is inappropriate to plot a line spectrum. If we regard $R_p^2/2$ as the contribution to variance in the range $\omega_p \pm \pi/N$, we can plot a histogram whose height in the range $\omega_p \pm \pi/N$ is such that

$R_p^2/2$ = area of histogram rectangle

= height of histogram x $2\pi/N$.

Thus the height of the histogram is given by

$$I(\omega_p) = NR_p^2/4\pi. \tag{7.16}$$

As usual, (7.16) does not apply for $p = N/2$; we may regard $a_{N/2}^2$ as the contribution to variance in the range $[\pi(N-1)/N,\pi]$ so that

$$I(\pi) = N\, a_{N/2}^2/\pi.$$

The plot of $I(\omega)$ against ω is usually called the *periodogram* even though $I(\omega)$ is a function of frequency rather than period. Other authors define the periodogram in a slightly different way, as some other multiple of R_p^2. For example Granger and Hatanaka (1964, equation 1.2.3) define a quantity which is π x expression (7.16). [Note that these authors use ω in equation 1.2.3 to denote period, while in the rest of their book ω denotes frequency.] Hannan (1970, equation 3.8) defines the periodogram in terms of complex numbers as

$$\frac{1}{2\pi N} \left| \sum_{t=1}^{N} x_t e^{it\omega} \right|^2$$

which is ½ x expression (7.16). Anderson (1971, Section 4.3.2) describes the graph of R_p^2 against the *period* N/p as the periodogram, and suggests the term *spectrogram* to describe the graph of R_p^2 against frequency. The advantage of definition (7.16) is that the total area under the periodogram is equal to the variance of the time series. Expression (7.16) may readily be calculated directly from the data by

$$I(\omega_p) = [(\Sigma x_t \cos 2\pi pt/N)^2 + (\Sigma x_t \sin 2\pi pt/N)^2]/N\pi.$$

$$(7.17)$$

Equation (7.17) also applies for $p = N/2$. Jenkins and Watts (1968) define a similar expression in terms of the variable $f = \omega/2\pi$, but call it the 'sample spectrum'.

The periodogram appears to be a natural way of estimating the power spectral density function, but we shall see that for a process with a *continuous* spectrum it provides a poor estimate and needs to be modified.

7.3.1 The relationship between the periodogram and the autocovariance function

The periodogram ordinate, $I(\omega)$, and the autocovariance coefficient, c_k, are both quadratic forms of the data $\{x_t\}$. It is of interest to see how they are related. We will show that the periodogram is the finite Fourier transform of $\{c_k\}$.

Using (7.2) we may rewrite (7.17) for $p \neq N/2$ as

$$I(\omega_p) = [(\Sigma(x_t - \bar{x}) \cos \omega_p t)^2 + (\Sigma(x_t - \bar{x}) \sin \omega_p t)^2]/N\pi$$

$$= \sum_{s,t=1}^{N} (x_t - \bar{x})(x_s - \bar{x})(\cos \omega_p t \cos \omega_p s$$

$$+ \sin \omega_p t \sin \omega_p s) /N\pi$$

But $\sum_{t=1}^{N-k} (x_t - \bar{x})(x_{t+k} - \bar{x})/N = c_k$ [see (4.1)]

and $\cos \omega_p t \cos \omega_p (t+k) + \sin \omega_p t \sin \omega_p (t+k) = \cos \omega_p k$, so that

$$I(\omega_p) = (c_0 + 2 \sum_{k=1}^{N-1} c_k \cos \omega_p k)/\pi \qquad (7.18)$$

$$= \sum_{k=-(N-1)}^{N-1} c_k e^{-i\omega_p k} /\pi. \qquad (7.19)$$

Equation (7.18) is perhaps the more useful relationship connecting the periodogram and autocovariance function. The equivalent form (7.19) is the more familiar form of a finite Fourier transform assuming that $c_k = 0$ for $|k| \geqslant N$.

7.3.2 Properties of the periodogram
When the periodogram is expressed in the form (7.18), it appears to be the 'obvious' estimate of the power spectrum

$$f(\omega) = (\gamma_0 + 2 \sum_{k=1}^{\infty} \gamma_k \cos \omega k)/\pi$$

simply replacing γ_k by its estimate c_k for values of k up to $(N - 1)$, and putting subsequent estimates of γ_k equal to zero. But although we find

$$\underset{N \to \infty}{E} (I(\omega)) \to f(\omega) \qquad (7.20)$$

so that the periodogram is asymptotically unbiased, we will see that the variance of $I(\omega)$ does *not* decrease as N increases. Thus $I(\omega)$ is *not a consistent* estimator for $f(\omega)$). An example of a periodogram is given by Kendall and Stuart (1966, p. 459) and the graph fluctuates wildly even though the series concerned is very long. This result is perhaps not too surprising when one realizes that the Fourier series representation (7.9) requires one to evaluate N parameters from N observations however long the series. Thus in Section 7.4 we will consider alternative ways of estimating a power spectrum which are essentially ways of *smoothing* the periodogram.

We complete this section by proving that $I(\omega)$ is not a consistent estimator for $f(\omega)$ in the case where $(x_1, \ldots, x_N)$ are taken from a discrete purely random process, where the observations are independent $N(\mu, \sigma^2)$ variates. It can also be shown that this result holds for other stationary processes with continuous spectra (e.g. Bartlett, 1966, p. 308) but this will not be demonstrated here.

From (7.10) we see that a_p and b_p are linear combinations of normally distributed random variables and so will themselves be normally distributed. Using (7.2)–(7.4), it can be shown (exercise 4) that a_p and b_p each have mean zero and variance $2\sigma^2/N$ for $p \neq N/2$. Furthermore we have

$$\text{Covariance}(a_p, b_p) = 4 \text{ Cov } \{(\Sigma x_t \cos \omega_p t), (\Sigma x_t \sin \omega_p t)\}/N^2$$

$$= 4 \sigma^2 (\Sigma \cos \omega_p t \sin \omega_p t)/N^2$$

since the $\{x_t\}$ are independent. Thus, using (7.5), we see that a_p and b_p are uncorrelated. Since (a_p, b_p) are bivariate normal, zero correlation implies that a_p and b_p are independent. Now a result from distribution theory says that if Y_1, Y_2 are independent $N(0, 1)$ variables, then $(Y_1^2 + Y_2^2)$ has a χ^2 distribution with 2 degrees of freedom, which is written χ_2^2. Thus

$$\frac{N(a_p^2 + b_p^2)}{2\sigma^2} = \frac{I(\omega_p) \times 2\pi}{\sigma^2}$$

is χ_2^2. Now the variance of a χ^2 distribution with ν degrees of freedom is 2ν, so that

$$\text{Var}[I(\omega_p) \times 2\pi/\sigma^2] = 4$$

and

$$\text{Var}[I(\omega_p)] = \sigma^4/\pi^2$$

As this variance is a constant, it does *not* tend to zero as $N \to \infty$, and hence $I(\omega_p)$ is not a consistent estimator for $f(\omega_p)$. Furthermore it can be shown that neighbouring periodogram ordinates are asymptotically independent, which further explains the very irregular form of an observed periodogram. Thus the periodogram needs to be modified in order to obtain a good estimate of a continuous spectrum.

7.4 Spectral analysis: some consistent estimation procedures

This section describes several alternative procedures for carrying out a spectral analysis. The different methods will be compared in Section 7.6. Each method provides a *consistent* estimate of the (power) spectral density function, in contrast to the periodogram. But although the periodogram is itself an inconsistent estimate, we shall see that the procedures described in this section are essentially based on the periodogram by using some sort of smoothing procedure.

Throughout the section we will assume that any obvious trend or seasonal variation has been removed from the data. If this is not done, the results of the spectral analysis are likely to be dominated by these effects, making any other effects difficult or impossible to see. Trend will produce a peak at zero frequency, while seasonal variation produces peaks at the seasonal frequency and at integer multiples of the seasonal frequency. These integer multiples of the fundamental frequency are called *harmonics* (see Section 7.8). Godfrey (1974) has pointed out that, for non-stationary series, it is often true that the final results depend more on the method chosen to remove trend than on any other factor.

Another type of non-stationarity is the presence of brief transient signals such as step changes. Brillinger (1973) has suggested a method of dealing with this problem.

7.4.1 Transforming the truncated autocovariance function

A popular type of estimation procedure consists of taking a Fourier transform of the truncated sample autocovariance function using a weighting procedure. From equation (7.18), we have that the periodogram is the discrete Fourier transform of the complete sample autocovariance function. But it is clear that the precision of the c_k decreases as k increases, so that it would seem intuitively reasonable to give less weight to the values of c_k as k increases. An estimator which has this property is:

$$\hat{f}(\omega) = \frac{1}{\pi} \left\{ \lambda_o c_o + 2 \sum_{k=1}^{M} \lambda_k c_k \cos \omega k \right\} \qquad (7.21)$$

where $\{\lambda_k\}$ are a set of weights called the *lag window*, and

$M(< N)$ is called the *truncation point.* Comparing (7.21) with (7.18) we see that values of c_k for $M < k < N$ are no longer used, while values of c_k for $k \leqslant M$ are weighted by a factor λ_k.

In order to use the above estimator, the reader must choose a suitable lag window and a suitable truncation point. The two lag windows which are in common use today are:

(a) *Tukey Window*

$$\lambda_k = \tfrac{1}{2}(1 + \cos \frac{\pi k}{M}) \quad k = 0, 1, \ldots M.$$

This window is also called the Tukey-Hanning or Blackman-Tukey window.

(b) *Parzen Window*

$$\lambda_k = \begin{cases} 1 - 6\left(\dfrac{k}{M}\right)^2 + 6\left(\dfrac{k}{M}\right)^3 & 0 \leqslant k \leqslant M/2 \\ 2(1 - k/M)^3 & M/2 \leqslant k \leqslant M. \end{cases}$$

These two windows are illustrated in Figure 7.1 with $M = 20$.

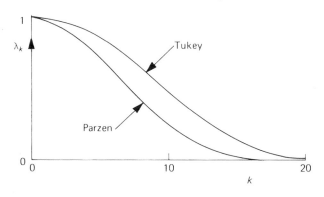

Figure 7.1   The Tukey and Parzen lag windows with
M = 20.

The Tukey and Parzen windows will give very much the same estimated spectrum for a given time series, although the Parzen window has a slight advantage in that it cannot give negative estimates. Many other lag windows have been suggested (see Hannan, 1970, Section 5.4) and 'window carpentry' was a popular research topic in the 1950's. Ways of comparing different windows will be discussed in Section 7.6. The well-known Bartlett window, with $\lambda_k = 1 - k/M$ for $k = 0, 1, \ldots, M$, is no longer used as its properties are inferior to the Tukey and Parzen windows. Recently Neave (1972b) has suggested a new window which would have superior properties but which is more complicated to use.

The choice of the truncation point, M, is rather difficult and little clear-cut advice is available in the literature. It has to be chosen subjectively so as to balance 'resolution' against 'variance'. The smaller the value of M, the smaller will be the variance of $\hat{f}(\omega)$ but the larger will be the bias (Neave, 1971). If M is too small, important features of $f(\omega)$ may be smoothed out, while if M is too large the behaviour of $\hat{f}(\omega)$ becomes more like that of the periodogram with erratic variation. Thus a compromise value must be chosen, usually in the range $1/20 < M/N < 1/3$. For example if $100 < N < 200$, a value of M about $N/6$ may be appropriate, while if $1000 < N < 2000$ a value of M less than $N/10$ may be appropriate. The asymptotic situation we have in mind is that as $N \to \infty$, so also does $M \to \infty$ but in such a way that $M/N \to 0$. A somewhat larger value of M is required for the Parzen window than for the Tukey window. Jenkins and Watts (1968) suggest trying 3 different values of M. A low value will give an idea where the large peaks in $f(\omega)$ are, but the curve is likely to be too smooth. A high value is likely to produce a curve showing a large number of peaks, some of which may be spurious. A compromise can then be achieved with the third value of M. As Hannan (1970, p. 311) says, 'experience is the real teacher and that cannot be got from a book'.

In principle (7.21) may be evaluated at any value of ω in $(0, \pi)$, but it is usually evaluated at equal intervals at $\omega = \pi j/Q$ for $j = 0, 1, \ldots, Q$, where Q is chosen sufficiently large to show up all features of $\hat{f}(\omega)$. Often Q is chosen to be equal to M.

The graph of $\hat{f}(\omega)$ against ω can then be plotted and examined.

7.4.2 Hanning This procedure, named after Julius Von Hann, is equivalent to the use of the Tukey window as described in Section 7.4.1, but adopts a different computational procedure. The estimated spectrum is calculated in two stages. Firstly a truncated unweighted cosine transform of the data is taken to give

$$\hat{f}_1(\omega) = \frac{1}{\pi} \left(c_0 + 2 \sum_{k=1}^{M} c_k \cos \omega k \right). \tag{7.22}$$

This is the same as (7.21) except that the lag window is taken to be unity (i.e. $\lambda_k = 1$). The estimates given by (7.22) are calculated at $\omega = \pi j/M$ for $j = 0, 1, \ldots, M$. These estimates are then smoothed using the weights ($\frac{1}{4}, \frac{1}{2}, \frac{1}{4}$) to give the Hanning estimates.

$$\hat{f}(\omega) = \tfrac{1}{4} \hat{f}_1(\omega - \pi/M) + \tfrac{1}{2} \hat{f}_1(\omega) + \tfrac{1}{4}\hat{f}_1(\omega + \pi/M) \tag{7.23}$$

at $\omega = \pi j/M$ for $j = 1, 2, \ldots, (M - 1)$. At zero frequency, and at the Nyquist frequency, π, we take

$$\hat{f}(0) = \tfrac{1}{2}[\hat{f}_1(0) + \hat{f}_1(\pi/M)]$$

$$\hat{f}(\pi) = \tfrac{1}{2}[\hat{f}_1(\pi) + \hat{f}_1(\pi(M - 1)/M)]$$

It is easily demonstrated that this procedure is equivalent to the use of the Tukey window. Substituting (7.22) into (7.23) we find

$$\hat{f}(\omega) = \frac{1}{\pi} \left\{ c_0 + 2 \sum_{k=1}^{M} c_k [\tfrac{1}{4} \cos (\omega - \pi/M)k + \tfrac{1}{2} \cos \omega k \right.$$
$$\left. + \tfrac{1}{4} \cos (\omega + \pi/M)k] \right\}$$

But $\cos(\omega - \pi/M)k + \cos(\omega + \pi/M)k = 2 \cos \omega k \cos(\pi k/M)$ and comparing with (7.21) we see that the lag window is indeed the Tukey window.

There is relatively little difference in the computational efficiency of Hanning and the straightforward use of the Tukey window. Both methods yield the same estimates and so it matters little which of the two procedures is used in practice.

7.4.3 Hamming This technique is very similar to Hanning and has a very similar title which sometimes leads to confusion. In fact Hamming is named after a quite different person, namely R. W. Hamming. The technique is nearly identical to Hanning except that the weights (¼, ½, ¼) in (7.23) are changed to (0.23, 0.54, 0.23). At the frequencies $\omega = 0$ and $\omega = \pi$, the weights are 0.54 and 0.46. The procedure gives similar estimates to those produced by Hanning.

7.4.4 Smoothing the periodogram The methods of Sections (7.4.1)–(7.4.3) are based on transforming the sample autocovariance function. An alternative type of approach is to smooth the periodogram by simply grouping the periodogram ordinates in sets of size m and finding their average value. This approach is based on a suggestion by P. J. Daniell in 1946. Then we find

$$\hat{f}(\omega) = \frac{1}{m} \sum_j I(\omega_j) \tag{7.24}$$

where $\omega_j = 2\pi j/N$ and j varies over m consecutive integers so that the ω_j are symmetric about ω. In order to estimate $f(\omega)$ at $\omega = 0$ and $\omega = \pi$, (7.24) has to be modified in an obvious way, treating the periodogram as being symmetric about 0 and π. Then, taking m to be odd with $m^* = (m - 1)/2$, we have

$$\hat{f}(0) = 2 \sum_{j=1}^{m^*} I(2\pi j/N)/m$$

143

assuming $I(0) = 0$. We also have

$$\hat{f}(\pi) = \left[I(\pi) + 2 \sum_{j=1}^{m} {}^{*} I(\pi - 2\pi j/N) \right] /m.$$

Now we know that the periodogram is asymptotically unbiased but inconsistent for the true spectrum. Since neighbouring periodogram ordinates are asymptotically uncorrelated, it is clear that the variance of (7.24) will be of order $1/m$. It is also clear that the estimator (7.24) may be biased since

$$E(\hat{f}(\omega)) \simeq \frac{1}{m} \sum_{j} f(\omega_j)$$

which is equal to $f(\omega)$ only if the spectrum is linear over the interval. However the bias will be unimportant provided that $f(\omega)$ is a reasonably smooth function at ω and m is not too large compared with N.

Thus the choice of m is rather like the choice of the truncation point, M, in Section 7.4.1 in that it has to be chosen so as to balance resolution against variance, although the effects are in opposite directions. The larger the value of m the smaller will be the variance of the resulting estimate but the larger will be the bias, and if m is too large then interesting features of $f(\omega)$, such as peaks, may be smoothed out. As N increases, so we can allow m to increase.

There seems to be relatively little advice in the literature on the choice of m. It seems advisable to try several values, in the region of $N/40$. A high value should give some idea where the large peaks in $f(\omega)$ are, but the curve is likely to be too smooth. A low value is likely to produce a curve showing a large number of peaks, some of which may be spurious. A compromise can then be made.

Although the procedure described in this section is computationally quite different from that of Section 7.4.1, there are in fact close links between the two procedures. In

Section 7.3.1 we derived the relationship between the periodogram and the sample autocovariance function, and if we substitute equation (7.18) into (7.24) we can express the estimate, $\hat{f}(\omega)$, in terms of the sample autocovariance function in a similar form to equation (7.21). We find, after some algebra (exercise 5), that the truncation point is $(N - 1)$ and the lag window is given by

$$\lambda_k = \begin{cases} 1 & k = 0 \\[2mm] \dfrac{\sin (m\pi k/N)}{m \sin (\pi k/N)} & k = 1, 2, \ldots, N-1 \end{cases}$$

Historically, the method was not much used until recent years because it apparently requires much more computational effort than the procedure of Section 7.4.1. Calculating the periodogram using equation (7.17) at ω_p for $p = 1, 2, \ldots, N/2$, would require about N^2 arithmetic operations (each one a multiplication and an addition), whereas using equation (7.21) less than MN operations are required to calculate the $\{c_k\}$ so that the total number of operations is only of order $M(N + M)$ if $Q = M$. Two factors have led to the increasing use of the smoothed periodogram. Firstly the advent of high speed computers means that it is no longer necessary to restrict oneself to the method requiring the fewest calculations (Jones, 1965). For example when $N = 1000$, it takes less than a minute to compute the periodogram on an IBM 7094 using equation (7.17). The second factor has been the rediscovery of a technique called the Fast Fourier Transform which can speed up the computation of the periodogram quite considerably. This technique will now be described.

7.4.5 The fast Fourier transform   This technique substantially reduces the time required to perform a Fourier analysis on a computer, and is also more accurate. The title is usually abbreviated to FFT and we will use this abbreviation. (But note that Hannan (1970) uses this abbreviation to denote *finite* Fourier transform.)

A history of the FFT is described by Cooley *et al.* (1967), the ideas going back to the early 1900's. But it was the work of J. W. Cooley, J. W. Tukey and G. Sande about 1965 which first stimulated the application of the technique to time-series analysis. Much of the subsequent work has been published in IEEE Transactions on Audio and Electroacoustics (now renamed IEEE Trans. on Acoustics, Speech and Signal Processing) which devoted two issues in June, 1967, and June, 1969, entirely to this topic. We will only give a broad outline of the technique here. For further details, see for example Otnes and Enochson (1972, Chapter 4), Bingham *et al.* (1967), and Bendat and Piersol (1971).

The basic idea of the FFT can be illustrated in the case when N can be factorized in the form $N = rs$. If we assume that N is even, then at least one of the factors, say r, will be even. Using complex numbers for mathematical simplicity, the Fourier coefficients from equation (7.10) are given by

$$a_p + ib_p = 2[\Sigma x_t \exp(2\pi ipt/N)]/N \qquad (7.25)$$

for $p = 0, 1, 2, \ldots, N/2 - 1$. For mathematical convenience we denote the observations by $x_0, x_1, \ldots, x_{N-1}$ so that the summation in (7.25) is from $t = 0$ to $(N - 1)$. Now we can write t in the form

$$t = rt_1 + t_0$$

where $t_1 = 0, 1, \ldots, (s - 1)$, and $t_0 = 0, 1, \ldots, (r - 1)$, as t goes from 0 to $(N - 1)$, in view of the fact that $N = rs$. Similarly we can decompose p in the form

$$p = sp_1 + p_0$$

where $p_1 = 0, 1, \ldots, (r/2 - 1)$, and $p_0 = 0, 1, \ldots, (s - 1)$, as p goes from 0 to $(N/2 - 1)$. Then the summation in (7.25) may be written

$$\sum_{t_0 = 0}^{r-1} \exp(2\pi i p t_0 / N) \sum_{t_1 = 0}^{s-1} x_t \exp(2\pi i p r t_1 / N).$$

But $\exp(2\pi i p r t_1 / N) = \exp(2\pi i (s p_1 + p_0) r t_1 / N) = \exp(2\pi i p_0 r t_1 / N)$

since $\exp(2\pi i s p_1 r t_1 / N) = \exp(2\pi i p_1 t_1) = 1$ for all p_1, t_1.

Thus $\sum_{t_1 = 0}^{s-1} x_t \exp(2\pi i p r t_1 / N)$

does not depend on p_1 and is therefore a function of t_0 and p_0 only, say $A(p_0, t_0)$. Then (7.25) may be written

$$a_p + i b_p = 2 \left[\sum_{t_0 = 0}^{r-1} A(p_0, t_0) \exp(2\pi i p t_0 / N) \right] \Big/ N.$$

Now there are rs functions of type $A(p_0, t_0)$ to be calculated, each requiring s complex multiplications and additions. Then the $(a_p + i b_p)$ may be calculated with $(r^2 s / 2)$ complex multiplications and additions giving a grand total of $rs(s + r/2) = N(s + r/2)$ calculations instead of the $N^2 / 2$ calculations required to use (7.25) directly.

Much bigger reductions in computing can be made by an extension of the above procedure when N is highly composite (i.e. has many small factors). In particular if N is of the form 2^n, then we find that the number of operations is of order Nn (or $N \log_2 N$) instead of N^2. Substantial gains can also be made when N has several factors (e.g. $N = 2^p 3^q 5^r \ldots$) and Singleton (1969) gives a general computer program.

In practice it is unlikely that N will be of a simple form such as 2^n, although it may be possible to make N highly composite by omitting a few observations. More generally we can add zeros to the (mean-corrected) sample record so as to *increase* the value of N until it is a suitable integer. This should be accompanied by a device called *tapering* (e.g. Tukey, 1967; Godfrey, 1974) to make the data smooth. Suppose for example that we have 382 observations. This value of N is not highly composite. We could proceed as follows.

(a) Remove any linear trend from the data, and keep the residuals (which should have mean zero) for subsequent analysis. If there is no trend, simply subtract the overall mean from each observation.

(b) Apply a linear taper to about 5% of the data at each end. In this example, if we denote the detrended mean-corrected data by $x_0, x_1, \ldots, x_{381}$, then the tapered series is given by

$$x_t{}^* = \begin{cases} (t+1)x_t/20 & t = 0, 1, \ldots, 18 \\ (382 - t)x_t/20 & t = 363, \ldots, 381 \\ x_t & t = 19, 20, \ldots, 362 \end{cases}$$

(c) Add $(512 - 382) = 130$ zeros at one end of the tapered series, so that $N = 512 = 2^9$.

(d) Carry out an FFT on the data, calculate the Fourier coefficients $a_p + ib_p$, and average the values of $(a_p^2 + b_p^2)$ in groups of about 10.

In fact with N as low as 382, the computational advantage of the FFT is limited and we could equally well calculate the periodogram directly which avoids the need for tapering and adding zeros. The FFT really comes into its own for $N > 1000$.

Jenkins and Watts (1968) give two reasons why they think the case for using the FFT in spectral analysis is not strong. Firstly they say that fast computers are more than adequate for carrying out a spectral analysis by traditional methods. This is certainly true for $N < 1000$, but with several thousand observations even fast computers will take several minutes without the FFT. Secondly Jenkins and Watts say that the autocorrelation function is an invaluable intermediate stage in spectral analysis. This is also true, but does not mean that the FFT is useless because it is generally quicker to calculate the sample autocovariance function by performing *two*

148

FFT's. As described by Bingham *et al.* (1967), we can compute the Fourier coefficients (a_p, b_p) with an FFT of the mean-corrected data at $\omega_p = 2\pi p/N$ for $p = 0, 1, \ldots, N - 1$ and not for $p = 0, 1, \ldots, N/2$ as we usually do. The extra coefficients are redundant for real-valued processes since $a_{N-k} = a_k$ and $b_{N-k} = -b_k$. We then compute $R_p^2 = a_p^2 + b_p^2$ and Fast Fourier retransform the sequence (R_p^2) to get the mean lagged products. With $N = 4096$, R. Fenton found that it took 700 seconds to calculate one sixth of the (c_k) directly on the Bath University computer but only 200 seconds to calculate *all* the (c_k) using two FFT's. In using the FFT for this purpose, one has to be careful to add enough zeros to the data (*without* tapering) to make sure that one gets non-circular sums of lagged products, as defined by equation (4.1) and used throughout this book. Circular coefficients result if zeros are not added where, for example, the circular autocovariance coefficient at lag 1 is

$$c_1^* = \left[\sum_{t=1}^{N} (x_t - \bar{x})(x_{t+1} - \bar{x}) \right]/N$$

where x_{N+1} is taken equal to x_1. If $x_1 = \bar{x}$, the circular and non-circular coefficients at lag 1 are the same. To calculate all the autocovariance coefficients of a set of N observations one adds N zeros, to make $2N$ 'observations' in all.

7.5 Confidence intervals for the spectrum

The methods of Section 7.4 all produce *point* estimates of the spectral density function at different frequencies. In this section we show how to find confidence intervals for the spectrum at different frequencies.

In Section 7.3.2 we showed that a white noise process, with constant spectrum $f(\omega) = \sigma^2/\pi$, has a periodogram ordinate at frequency ω, $I(\omega)$, which is such that $2I(\omega)/f(\omega)$ is distributed as χ_2^2. A more general result is given by Jenkins and Watts (1968, Section 6.4.2) for the estimator of

Section 7.4.1, namely

$$\hat{f}(\omega) = \left[\sum_{k=-M}^{M} \lambda_k c_k \cos \omega k \right] / \pi \,,$$

which is that asymptotically $\nu \hat{f}(\omega)/f(\omega)$ is approximately distributed as χ_ν^2, where

$$\nu = 2N \left/ \sum_{k=-M}^{M} \lambda_k^2 \right. \tag{7.26}$$

is called the number of degrees of freedom of the lag window. Then

$$P(\chi_{\nu,1-\alpha/2}^2 < \nu \hat{f}(\omega)/f(\omega) < \chi_{\nu,\alpha/2}^2) = 1 - \alpha$$

so that the $100(1 - \alpha)\%$ confidence interval for $f(\omega)$ is given by

$$\frac{\nu \hat{f}(\omega)}{\chi_{\nu,\alpha/2}^2} \text{ to } \frac{\nu \hat{f}(\omega)}{\chi_{\nu,1-\alpha/2}^2} \,.$$

The degrees of freedom for the Tukey and Parzen windows turn out to be $2.67N/M$ and $3.71N/M$ respectively. When smoothing the periodogram in groups of size m, it is clear that the result will have $2m$ degrees of freedom and there is no need to apply equation (7.26). In fact equation (7.26) does not work for the periodogram when expressed in the form (7.18) as noted by Hannan (1970, p. 281).

The confidence intervals given in this section are asymptotic results. Neave (1972a) has shown that these results are also quite accurate for short series.

7.6 A comparison of different estimation procedures

Several factors need to be considered when comparing the different estimation procedures which were described in Section 7.4. These include such practical considerations as computing time and the availability of computer programs.

We begin by considering the theoretical properties of the different procedures. Other comparative discussions are given by Jones (1965), Bartlett (1967), Neave (1972b) and Jenkins and Watts (1968).

It is useful to introduce a function called the *spectral window* or *kernel* which is the Fourier transform of the lag window. If we define the lag window, λ_k, to be zero for $k > M$, and to be symmetric so that $\lambda_{-k} = \lambda_k$, then the spectral window is given by

$$K(\omega) = \frac{1}{2\pi} \sum_{k=-\infty}^{\infty} \lambda_k \, e^{-ik\omega} \qquad (7.27)$$

for $(-\pi < \omega < \pi)$. This has the inverse relation

$$\lambda_k = \int_{-\pi}^{\pi} K(\omega) \, e^{i\omega k} \, d\omega. \qquad (7.28)$$

All the estimation procedures for the spectrum can be put in the general form

$$\hat{f}(\omega_0) = \frac{1}{\pi} \sum_{k=-N+1}^{N-1} \lambda_k \, c_k \, e^{-i\omega_0 k}$$

$$= \frac{1}{\pi} \sum \left[\int_{-\pi}^{\pi} K(\omega) e^{i\omega k} \, d\omega \right] c_k \, e^{-i\omega_0 k}$$

$$= \frac{1}{\pi} \int_{-\pi}^{\pi} K(\omega) \, [\Sigma \, c_k \, e^{ik(\omega - \omega_0)}] \, d\omega$$

$$= \int_{-\pi}^{\pi} K(\omega) \, I(\omega_0 - \omega) d\omega \qquad (7.29)$$

using equation (7.19). Equation (7.29) shows that all the estimation procedures are essentially smoothing the periodogram using the weight function $K(\omega)$. The value of the lag window at lag zero is usually specified to be one, so

151

that from (7.28) we have

$$\lambda_0 = 1 = \int_{-\pi}^{\pi} K(\omega)d\omega$$

which is a desirable property for a smoothing function.

Taking expectations in equation (7.29) we have asymptotically that

$$E[\hat{f}(\omega_0)] = \int_{-\pi}^{\pi} K(\omega) f(\omega_0 - \omega)d\omega. \qquad (7.30)$$

Thus the spectral window is a weight function expressing the contribution of the spectral density function at each frequency to the expectation of $\hat{f}(\omega_0)$. The name 'window' arises from the fact that $K(\omega)$ determines the part of the periodogram which is 'seen' by the estimator.

Examples of the spectral windows for the three commonest methods of spectral analysis are shown in Fig. 7.2, which is adapted from Jones (1965, Fig. 5). Taking $N = 1000$, the spectral window for the smoothed periodogram with $m = 20$ is shown as line A. The other two windows are the Parzen and Tukey windows, denoted by lines B and C. The values of the truncation point, M, were chosen to be 93 for the Parzen window and 67 for the Tukey window. These values of M were choosen so that all three windows gave estimators with equal variance. Formulae for variances will be given later in this section.

Inspecting Fig. 7.2, we see that the Parzen and Tukey windows look very similar, although the Parzen window has the advantage of being non-negative and of having smaller side-lobes. The shape of the periodogram window is quite different. It is approximately rectangular with a sharp cut-off and is close to the 'ideal' band-pass filter which would be exactly rectangular but which is unattainable in practice. The periodogram window also has the advantage of being non-negative. Overall we can say that the periodogram has a

152

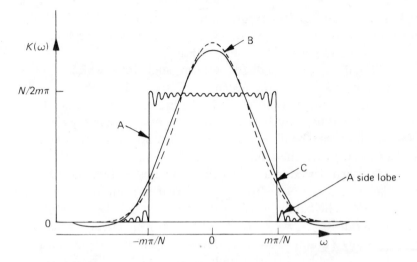

Figure 7.2   The spectral windows for three common methods of spectral analysis; A: Smoothed periodogram (m = 20); B: Parzen (M = 93); C: Tukey (M = 67); all with N = 1000.

window with more desirable properties than those of Parzen and Tukey.

In comparing different windows, we also want to consider *bias, variance* and *bandwidth.* We will not derive formulae for the bias produced by the different procedures. It is clear from equation (7.30) that the wider the window, the larger will be the bias. In particular it is clear that all the smoothing procedures will tend to lower peaks and raise troughs.

As regards variance, we have from Section 7.5 that $v\hat{f}(\omega)/f(\omega)$ is approximately distributed as χ^2_v, where $v = 2m$, $3.71 \, N/M$, and $8N/3M$ for the smoothed periodogram, Parzen and Tukey windows, respectively. But

$$\text{Var}(\chi^2_v) = 2v$$

and

$$\text{Var}(v\hat{f}(\omega)/f(\omega)) = v^2 \, \text{Var}(\hat{f}(\omega)/f(\omega))$$

153

and so $\text{Var}(\hat{f}(\omega)/f(\omega))$ turns out to be $1/m$, $2M/3.71\ N$, and $3M/4N$ for the three windows. Equating these expressions gives the values of M chosen for Fig. 7.2.

Finally let us introduce the term bandwidth, which roughly speaking is the width of the spectral window. Various definitions are given in the literature, but we shall adopt that used by Jenkins and Watts (1968), namely the width of the 'ideal' rectangular window which would give an estimator with the same variance. The window of the smoothed periodogram is so close to being rectangular for m 'large' that it is clear from Fig. 7.2 that the bandwidth will be $2\pi m/N$. The bandwidths for the Parzen and Tukey windows turn out to be $2\pi \times 1.86/M$ and $8\pi/3M$ respectively. When plotting a graph of an estimated spectrum, it is a good idea to indicate the bandwidth which has been used.

The choice of bandwidth, or equivalently the choice of m or M, is an important step in spectral analysis. For the Parzen and Tukey windows, the bandwidth is inversely proportional to M (see Fig. 7.3). As M gets larger, the window gets narrower and the bias gets smaller but the variance of the resulting estimator gets larger. In fact the variance is inversely proportional to the bandwidth. For the smoothed periodogram, the bandwidth is directly proportional to m.

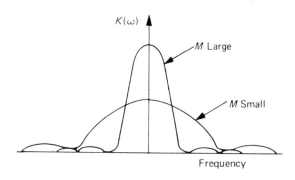

Figure 7.3   Spectral windows for different values of M.

For the unsmoothed periodogram, with $m = 1$, the window is very tall and narrow giving an estimator with large variance as we have already shown. All in all, the choice of bandwidth is rather like the choice of class interval when constructing a histogram.

We will now summarize the comparative merits of the three main estimation procedures. From the preceding discussion, it appears that the smoothed periodogram has superior theoretical properties in that its spectral window is approximately rectangular. From a computing point of view, it can be much slower for large N unless the FFT is used. If the FFT is used however, it can be much quicker and it is also possible to calculate the autocorrelation function using the FFT. For small N, computing time is relatively unimportant. Regarding computer programs, it is much easier to write a program for the Parzen or Tukey windows, but programs for the FFT are becoming more readily available and this trend will no doubt continue. In the future the use of the smoothed periodogram will probably become more general.

7.7 Analysing a continuous time series

We have so far been concerned with the spectral analysis of discrete time series. But time series are sometimes recorded as a continuous trace, as for example air temperature, the moisture content of tobacco emerging from a processing plant, and humidity. For series which contain components at very high frequencies, such as those arising in acoustics and speech processing, it may be possible to analyse them mechanically using tuned filters, but the more usual procedure is to *digitize* the series by reading off the values of the trace at discrete intervals. If values are taken at equal time intervals of length Δt, we have converted a continuous time series into a standard discrete time series and can use the methods already described.

In sampling a continuous time series, the main question is how to choose the sampling interval Δt. It is clear that sampling leads to some loss of information and that this loss gets worse as Δt increases. However it is expensive to make Δt very small and so a compromise value must be sought.

For the sampled series, the Nyquist frequency is $\pi/\Delta t$ radians per unit time, and we can get no information about variation at higher frequencies. Thus we clearly want to choose Δt so that variation in the continuous series is negligible at frequencies higher than $\pi/\Delta t$. In fact most measuring instruments are *band-limited* in that they do not respond to frequencies higher than a certain maximum frequency. If this maximum frequency is known or can be guessed, then the choice of Δt is straightforward.

If Δt is chosen to be too large, then a phenomenon called *aliasing* may occur. This can be illustrated by the following theorem.

Theorem 7.1   A continuous time series, with spectrum $f_c(\omega)$ for $0 < \omega < \infty$, is sampled at equal intervals of length Δt. The resulting discrete time series has spectrum $f_d(\omega)$ defined over $0 < \omega < \pi/\Delta t$. Then $f_d(\omega)$ and $f_c(\omega)$ are related by

$$f_d(\omega) = \sum_{s=0}^{\infty} f_c(\omega + 2\pi s/\Delta t) + \sum_{s=1}^{\infty} f_c(-\omega + 2\pi s/\Delta t) \quad (7.31)$$

Proof   The proof will be given for the case $\Delta t = 1$. The extension to other values of Δt is straightforward.

The acv.f's of the continuous and sampled series will be denoted by $\gamma(\tau)$ and γ_k. It is clear that when τ takes an integer value, say k, then

$$\gamma(k) = \gamma_k. \quad (7.32)$$

Now from (6.9) we have

$$\gamma_k = \int_0^{\pi} f_d(\omega) \cos \omega k \, d\omega$$

156

while from (6.18) we have

$$\gamma(\tau) = \int_0^\infty f_c(\omega) \cos \omega\tau \, d\omega.$$

Thus, using (7.32), we have

$$\int_0^\pi f_d(\omega) \cos \omega k \, d\omega = \int_0^\infty f_c(\omega) \cos \omega k \, d\omega$$

for $k = 0, \pm 1, \pm 2, \ldots$. Now

$$\int_0^\infty f_c(\omega) \cos \omega k \, d\omega = \sum_{s=0}^\infty \int_{2\pi s}^{2\pi(s+1)} f_c(\omega) \cos \omega k \, d\omega$$

$$= \sum_{s=0}^\infty \int_0^{2\pi} f_c(\omega + 2\pi s) \cos \omega k \, d\omega.$$

$$= \sum_{s=0}^\infty \int_0^\pi \{f_c(\omega + 2\pi s) + f_c[2\pi(s+1) - \omega]\} \cos \omega k \, d\omega$$

$$= \int_0^\pi \left\{ \sum_{s=0}^\infty f_c(\omega + 2\pi s) + \sum_{s=1}^\infty f_c(2\pi s - \omega) \right\} \cos \omega k \, d\omega,$$

and the result follows.

The implications of this theorem are now considered. Firstly, we note that if the continuous series contains no variation at frequencies above the Nyquist frequency, so that $f_c(\omega) = 0$ for $\omega > \pi/\Delta t$, then $f_d(\omega) = f_c(\omega)$. In this case no information is lost by sampling. But more generally, the effect of sampling will be that variation at frequencies above the Nyquist frequency will be 'folded back' and produce an effect at a frequency lower than the Nyquist frequency in $f_d(\omega)$. If we denote the Nyquist frequency, $\pi/\Delta t$, by ω_N, then the frequencies ω, $2\omega_N - \omega$, $2\omega_N + \omega$, $4\omega_N - \omega$, ..., are called *aliases* of one another. Variation at all these frequencies in the continuous series will appear as variation at frequency ω in the sampled series.

From a practical point of view, aliasing will cause trouble

157

unless Δt is chosen so that $f_c(\omega) \simeq 0$ for $\omega > \pi/\Delta t$. If we have no advance knowledge about $f_c(\omega)$ then we can guesstimate a value for Δt. If the resulting estimate of $f_d(\omega)$ approaches zero near the Nyquist frequency $\pi/\Delta t$, then our choice of Δt was almost certainly sufficiently small. But if $f_d(\omega)$ does not approach zero near the Nyquist frequency, then it is probably wise to try a smaller value of Δt. Alternatively one can filter the continuous series to remove the high frequency components if one is interested in the low frequency components.

7.8 Discussion

Spectral analysis can be a useful exploratory diagnostic tool in the analysis of many types of time series. In this section we discuss how the estimated spectrum should be interpreted, when it is likely to be most useful and when it is likely to be least useful. We also discuss some of the practical problems arising in spectral analysis.

We begin this discussion with an example to give the reader some feel for the sorts of spectrum shape that may arise. Fig. 7.4 shows four sections of trace, labelled A, B, C and D, which were produced by four different processes. The figure also shows the corresponding long-run spectra, labelled J, K, L and M, but these are given in *random* order. Note that the four traces use the same scale, the length produced in one second being shown on trace D. The four spectra are plotted using the same linear scales. The peak in spectrum L is at 15 cycles per second (or 15 Hz). The reader is invited to decide which series goes with which spectrum before reading on.

Trace A is much smoother than the other three traces. Its spectrum is therefore concentrated at low frequency and is actually spectrum M. The other three spectra are much harder to distinguish. Trace B is somewhat smoother than C or D and corresponds to spectrum K which 'cuts off' at a lower frequency than J or L. Trace C corresponds to

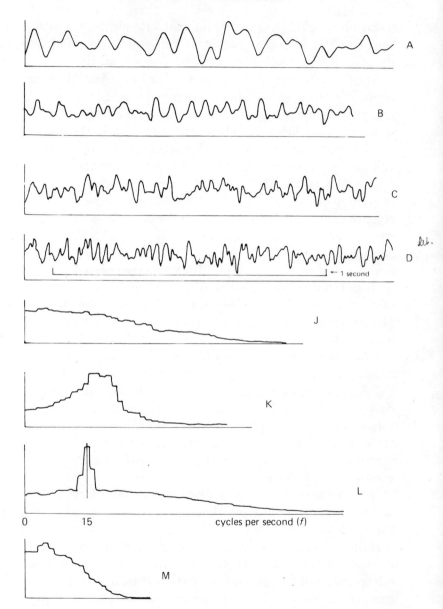

Figure 7.4   Four time series and their spectra. The spectra are given in random order.

159

spectrum J, while trace D contains a deterministic sinusoidal component at 15 cycles per second which contributes 20% of the total power. Thus D corresponds to spectrum L.

From a visual inspection of traces C and D, it is difficult or impossible to decide which goes with spectrum L. For this type of data, spectral analysis is invaluable in assessing the frequency properties. The reader may find it surprising that the deterministic component in trace D is so hard to see. In contrast the regular seasonal variation in air temperature at Recife given in Fig. 1.2 is quite obvious from a visual inspection of the time plot, but there the deterministic component accounts for 85% of the total variation. A spectral analysis of air temperature at Recife yields the spectrum shown in Fig. 7.5(a) with a large peak at a frequency of one cycle per year. But here the spectral analysis is not really necessary as the seasonal effect is obvious anyway. In fact if one has a series containing an obvious trend or seasonality, then such variation should be removed from the data *before* carrying out a spectral analysis, as any other effects will be relatively small and are unlikely to be visible in the spectrum of the raw data. Fig. 7.5(b) shows the spectrum of the Recife air temperature data when the seasonal variation has been removed. The variance is concentrated at low frequencies indicating either a trend, which is not apparent in Fig. 1.2, or short-term correlation as in a first-order AR process with a positive coefficient (c.f. Fig. 6.4a).

Removing trend and seasonality is the simplest form of a general procedure called *prewhitening*. It is easier to estimate the spectrum of a series which has a relatively flat spectrum. Prewhitening consists of making a linear transformation of the raw data so as to achieve a smoother spectrum, estimating the spectrum of the transformed data, and then using the transfer function of the linear transformation to estimate the spectrum of the raw data (see Chapter 9 and Anderson, 1971, p. 546). But this procedure requires some prior knowledge of

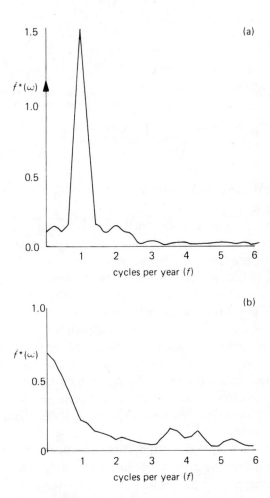

Figure 7.5   The estimated spectrum for monthly observations on air temperature at Recife; (a) for the raw data; (b) for the seasonally adjusted data.

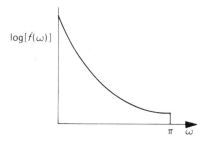

*Figure 7.6   The typical spectral
shape of an economic time-series.*

the spectral shape and is not often used except for removing
trend and seasonality.

Having estimated the spectrum of a given time series, how
do we interpret the results? There are various features to look
for. Firstly, are there any peaks in the spectrum? If so, why?
Secondly, is the spectrum large at low frequency, indicating
possible non-stationarity in the mean? Thirdly what is the
general shape of the spectrum? The typical shape of the power
spectrum of an economic variable is shown in Fig. 7.6, and
the implications of this shape are discussed by Granger
(1966). There are exceptions. Granger and Hughes (1971)
found a peak at a frequency of 1 cycle per 13 years when
they analysed Beveridge's yearly wheat price index series. But
this series covers 369 years which is much longer than most
economic series.

The general shape of the spectrum may occasionally be
helpful in indicating an appropriate parametric model, but it
is not generally used in this way. The spectrum is not, for
example, used in the Box—Jenkins procedure for identifying
an appropriate ARIMA process (though Hannan, 1970,
Section 6.5 has suggested that it might be). Spectral analysis
is essentially a non-parametric procedure in which a finite set
of observations is used to estimate a function defined over
the range $(0, \pi)$. The function is not constrained to any

particular parametric class and so spectral analysis is a more
general procedure than inference based on a particular
parametric class of models, but is also likely to be less
accurate if a parametric model really is appropriate.

Spectral analysis is at its most useful for series of the type
shown in Fig. 7.4, with no obvious trend or 'seasonal'
variation. Such series arise mostly in the physical sciences. In
economics, spectral techniques have perhaps not proved as
useful as was first hoped, although there have been some
successes. Recently, attempts have been made to apply
spectral analysis to marketing data but it can be argued
(Chatfield, 1974) that marketing series are usually too short
and the seasonal variation too large for spectral analysis to
give useful results. In meteorology and oceanography,
spectral analysis can be very useful (e.g. Craddock, 1965;
Munk *et al.*, 1959; Mooers and Smith, 1968; Snodgrass *et al.*,
1966; Roden, 1966) but even in these sciences, spectral
analysis often produces no worthwhile results. Chatfield and
Pepper (1970) analysed a number of monthly geophysical
series but found no tendency to oscillate at frequencies other
than the obvious annual effect.

We conclude this section by commenting on some practical
aspects of spectral analysis.

Most aspects, such as the choice of truncation point, have
already been discussed and will be further clarified in the
example given in Section 7.9. One problem, which has not
been discussed, is whether to plot the estimated spectrum or
its logarithm. An advantage of plotting the estimated
spectrum on a logarithmic scale is that its asymptotic
variance is then independent of the level of the spectrum so
that confidence intervals for the spectrum are of constant
width on a logarithmic scale. For spectra showing large
variations in power, a logarithmic scale also makes it possible
to show more detail over a wide range. (For example in
measuring sound, engineers use decibels which are measured
on a logarithmic scale.) Jenkins and Watts (1968. p. 266)

163

suggest that spectrum estimates should always be plotted on a logarithmic scale. But Anderson (1971, p. 547) points out that this exaggerates the visual effects of variations where the spectrum is small. Thus it is often easier to interpret a spectrum plotted on an arithmetic scale as the area under the graph corresponds to power and one can more readily assess the importance of different peaks. So, while it is often useful to plot $\hat{f}(\omega)$ on a logarithmic scale in the initial stages of a spectral analysis, when trying different truncation points and testing the significance of peaks, this writer generally prefers to plot the estimated spectrum on an arithmetic scale in order to interpret the final result. It is also generally easier to interpret a spectrum if the frequency scale is measured in cycles per unit time (f) rather than radians per unit time (ω). This has been done in Figs. 7.4 and 7.5. A linear transformation of frequency does not affect the *relative* heights of the spectrum at different frequencies which are of prime importance, though it does change the absolute heights by a constant multiple.

Another point worth mentioning is the possible presence in estimated spectra of *harmonics*. When a spectrum has a large peak at some frequency ω, then related peaks may occur at $2\omega, 3\omega, \ldots$. These multiples of the fundamental frequency are called harmonics and generally speaking simply indicate the non-sinusoidal character of the main cyclical component. For example Mackay (1973) studied the incidence of trips to supermarkets by consumers and found (not surprisingly!) a basic weekly pattern with harmonics at two and three cycles per week.

Finally, a question that is often asked is how large a value of N is required to get a reasonable estimate of the spectrum. It is often recommended that between 100 and 200 observations is the minimum. Granger and Hughes (1968) have tried smaller values of N and conclude that only very large peaks can then be found. If the data is prewhitened, so that the spectrum is fairly flat, then reasonable estimates may be obtained even with values of N less than 100.

7.9 An example

As an example, we analysed part of trace D of Fig. 7.4. Although a fairly long trace was available, we decided just to analyse a section lasting for one second to illustrate the problems of analysing a fairly short series. This set of data will also illustrate the problems of analysing a continuous trace as opposed to a discrete series.

The first problem was to digitize the data, and this required the choice of a suitable sampling interval. Inspection of the original trace showed that variation seemed to be 'fairly smooth' over a length of 1 mm corresponding to $1/100$ second, but to ensure that there was no aliasing, we chose $1/200$ second as the sampling interval, giving $N = 200$ observations.

For such a short series, there is little to be gained by using the Fast Fourier Transform. We decided to transform the truncated autocovariance function, using equation (7.21), with the Tukey window. Several truncation points were tried, and the results for $M = 20$, 40 and 80 are shown in Fig. 7.7. Above about 70 cycles per second, the estimates produced by the three values of M are very close to one another when the spectrum is plotted on an arithmetic scale and cannot be distinguished on the graph. Equation (7.21) was evaluated at 51 points, taking $Q = 50$, at $\omega = \pi j/50$ ($j = 0, 1, \ldots, 50$) where ω is measured in radians per unit time. Now in this example 'unit time' is $1/200$ second and so the values of ω in radians per second are $\omega = 200\pi j/50$ for $j = 0, 1, \ldots, 50$. Thus the frequencies expressed in cycles per second, by $f = \omega/2\pi$, are $f = 2j$ for $j = 0, 1, \ldots, 50$. The Nyquist frequency is given by $f_N = 100$ cycles per second, which completes one cycle every two observations.

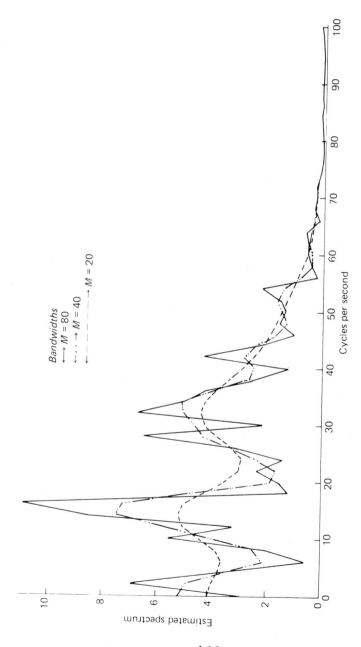

Figure 7.7 Estimated spectra for graph D of Fig. 7.4 using the Tukey window with (a) M = 80; (b) M = 40; (c) M = 20.

Looking at Fig. 7.7, the estimated spectrum is judged rather too smooth when $M = 20$, and much too erratic when $M = 80$. The value $M = 40$ looks about right although $M = 30$ might be even better. There is a clear peak at about 15 cycles per second (15 Hz) as there is in Fig. 7.4 (L), but there is also a smaller unexpected peak around 30 cycles per second, which looks like a harmonic of the variation at 15 cycles per second. If a longer series of observations were to be analysed, this peak might disappear indicating that the peak is spurious.

We also estimated the spectrum using a Parzen window with a truncation point of $M = 56$. This value was chosen so that the degrees of freedom of the window, namely 13.3, was almost the same as for the Tukey window with $M = 40$. I intended to plot the results on Fig. 7.7, but the graph was so close to that produced by the Tukey window that it was impossible to draw them both on the same graph. The biggest difference in estimates at the same frequency was 0.33 at 12 cycles per second but most of the estimates differed only in the second decimal place. It is clear that the Tukey and Parzen windows give very much the same estimates when equivalent values of M are used.

One feature to note in Fig. 7.7 is that the estimated spectrum approaches zero as the frequency approaches the Nyquist frequency. This suggests that there is no aliasing and that our choice of sampling interval was sufficiently small.

Also note that the bandwidths are indicated in Fig. 7.7. The bandwidth for the Tukey window is $8\pi/3M$ in radians per unit time. As 'unit time' is $1/200$ second, the bandwidth is $1600\pi/3M$ in radians per second or $800/3M$ in cycles per second.

Confidence intervals can be calculated as described in Section 7.5. For a sample of only 200 observations, they turn out to be disturbingly wide. For example when $M = 40$, the degrees of freedom are $2.67N/M = 13.3$. For convenience this is rounded off to the nearest integer, namely $\nu = 13$. The peak in the estimated spectrum is at 14 cycles per second where

$\hat{f}(\omega) = 7.5$. Here the 95% confidence interval is (3.9 to 19.5). Clearly a longer series is desirable to make the confidence intervals acceptably narrow.

Exercises

1. Revision of Fourier series. Show that the Fourier series which represents the function

$$f(x) = x^2 \qquad \text{in } -\pi \leqslant x \leqslant \pi$$

   is given by

$$f(x) = \frac{\pi^2}{3} - 4 \left[\frac{\cos x}{1} - \frac{\cos 2x}{2^2} + \frac{\cos 3x}{3^2} - \dots \right]$$

2. Derive equations (7.6) and (7.8).

3. Derive Parseval's theorem given by equation (7.15).

4. If $X_1, \dots, X_N$ are independent $N(\mu, \sigma^2)$ variates, show that

$$a_p = 2[\Sigma X_t \cos(2\pi pt/N)]/N$$

   is $N(0, 2\sigma^2/N)$ for $p = 1, 2, \dots, N/2 - 1$.

5. Derive the lag window for smoothing the periodogram in sets of size m. For algebraic simplicity take m odd, with $m = 2m^* + 1$, so that

$$\hat{f}(\omega_p) = \frac{1}{m} \sum_{j=-m^*}^{m^*} I(\omega_p + \frac{2\pi j}{N}).$$

6. Evaluate the degrees of freedom for the Tukey window using equation (7.26).

CHAPTER 8

Bivariate Processes

Thus far, we have been concerned with analysing a single time series. We now turn our attention to the situation where we have observations on *two* time-series and we are interested in the relationship between them.

Jenkins and Watts (1968, p. 322) distinguish two types of situation. In the first type of situation, the two series arise 'on an equal footing' and we are interested in the correlation between them. For example it is often of interest to analyse seismic signals received at two recording sites. In the second, more important type of situation, the two series are 'causally related' and one series is regarded as the *input* to a *linear system*, while the other series is regarded as the *output*. We are then interested in finding the properties of the linear system. The two types of situation are roughly speaking the time series analogues of *correlation* and *regression*.

The first type of situation is considered in this chapter where the cross-correlation function and the cross-spectrum are introduced. These functions are also useful in the study of linear systems which are discussed in Chapter 9.

8.1 Cross-covariance and cross-correlation functions

Suppose we have N observations on two variables, x and y, at unit time intervals over the same period. The observations will be denoted by $(x_1, y_1), \ldots, (x_N, y_N)$. These observations may be regarded as a finite realization of a discrete bivariate stochastic process (X_t, Y_t).

In order to describe a bivariate process it is useful to know the moments up to second order. For a univariate process, the moments up to second order are the mean and autocovariance function. For a bivariate process, the moments up to second order consist of the mean and autovariance functions for each of the two components plus a new function, called the *cross-covariance function*, which is given by

$$\gamma_{xy}(t, \tau) = \text{Cov}(X_t, Y_{t+\tau}).$$

We will only consider bivariate processes which are second-order stationary, so that all moments up to second order do not change with time. We will use the following notation

$$E(X_t) = \mu_x. \quad E(Y_t) = \mu_y.$$
$$\text{Cov}(X_t, X_{t+\tau}) = \gamma_{xx}(\tau)$$
$$\text{Cov}(Y_t, Y_{t+\tau}) = \gamma_{yy}(\tau)$$
$$\text{Cov}(X_t, Y_{t+\tau}) = \gamma_{xy}(\tau) \tag{8.1}$$

Note that some authors (e.g. Granger and Hatanaka, 1964) define the cross-covariance function by

$$\text{Cov}(X_t, Y_{t-\tau}) = \gamma_{xy}^*(\tau).$$

Comparing with (8.1) we see that

$$\gamma_{xy}(\tau) = \gamma_{xy}^*(-\tau).$$

The cross-covariance function differs from the autocovariance function in that it is *not* an even function, since

$$\gamma_{xy}(\tau) \neq \gamma_{xy}(-\tau).$$

Instead we have the relationship

$$\gamma_{xy}(\tau) = \gamma_{yx}(-\tau)$$

where the subscripts are reversed.

The size of the cross-covariance coefficients depends on

the units in which $X(t)$ and $Y(t)$ are measured. Thus for interpretative purposes, it is useful to standardize the cross-covariance function to produce a function called the cross-correlation function, $\rho_{xy}(\tau)$, which is defined by

$$\rho_{xy}(\tau) = \gamma_{xy}(\tau)/\sqrt{[\gamma_{xx}(0)\gamma_{yy}(0)]} \tag{8.2}$$

This function measures the correlation between $X(t)$ and $Y(t + \tau)$ and has the properties

(a) $\rho_{xy}(\tau) = \rho_{yx}(-\tau)$
(b) $|\rho_{xy}(\tau)| \leqslant 1$ (see exercise 2).

Whereas $\rho_{xx}(0)$, $\rho_{yy}(0)$ are both equal to one, the value of $\rho_{xy}(0)$ is *not* necessarily equal to one, a fact which is often overlooked. Equation 50.107 in Kendall and Stuart (1966) and equation 10.4 in Kendall (1973) are both incorrect for this reason.

8.1.1 Examples
Before discussing the estimation of cross-covariance and cross-correlation functions, we will derive the theoretical functions for two examples of bivariate processes. The first example is rather 'artificial', but the model in example 2 can be useful in practice.

Example I   Suppose that $\{X_t\}$, $\{Y_t\}$, are both formed from the same purely random process $\{Z_t\}$ which has mean zero, variance σ_Z^2, by

$$X_t = Z_t$$
$$Y_t = 0.5\, Z_{t-1} + 0.5\, Z_{t-2}.$$

Then using 8.1 we have

$$\gamma_{xy}(\tau) = \begin{cases} 0.5\, \sigma_Z^2 & \tau = 1, 2 \\ 0 & \text{otherwise} \end{cases}$$

171

Now the variances of the two components are given by

$$\gamma_{xx}(0) = \sigma_Z^2$$

and

$$\gamma_{yy}(0) = \sigma_Z^2/2$$

so that, using (8.2), we have

$$\rho_{xy}(\tau) = \begin{cases} 0.5\sqrt{2} & \tau = 1, 2 \\ 0 & \text{otherwise.} \end{cases}$$

Example 2. Suppose that

$$X_t = Z_{1,t}$$
$$Y_t = X_{t-d} + Z_{2,t} \qquad\qquad (8.3)$$

where $\{Z_{1,t}\}$, $\{Z_{2,t}\}$ are uncorrelated purely random processes with mean zero and variance σ_Z^2, and where d is an integer. Then we find

$$\gamma_{xy}(\tau) = \begin{cases} \sigma_Z^2 & \tau = d \\ 0 & \text{otherwise} \end{cases}$$

$$\rho_{xy}(\tau) = \begin{cases} 1/\sqrt{2} & \tau = d \\ 0 & \text{otherwise.} \end{cases}$$

In Chapter 9 we will see that equation (8.3) corresponds to putting noise into a linear system which consists of a simple delay of lag d and then adding more noise. The cross-correlation function has a peak at lag d corresponding to the delay in the system, a result which the reader should find intuitively reasonable.

8.1.2 Estimation

The 'obvious' way of estimating the cross-covariance and cross-correlation functions is by means of corresponding sample functions. With N pairs of observations $\{(x_i, y_i);$

$i = 1$ to N}, the sample cross-covariance function is

$$c_{xy}(k) = \begin{cases} \sum_{t=1}^{N-k} (x_t - \bar{x})(y_{t+k} - \bar{y})/N & [k = 0, 1, \ldots (N-1)] \\ \sum_{t=1-k}^{N} (x_t - \bar{x})(y_{t+k} - \bar{y})/N & \\ & [k = -1, -2, \ldots, -(N-1)] \end{cases}$$

(8.4)

and the sample cross-correlation function is

$$r_{xy}(k) = c_{xy}(k)/\sqrt{[c_{xx}(0)c_{yy}(0)]} \qquad (8.5)$$

where $c_{xx}(0)$, $c_{yy}(0)$ are the sample variances of observations on x_t and y_t respectively.

It can be shown that these estimators are asymptotically unbiased and consistent. However it can also be shown that successive estimates are themselves autocorrelated. In addition the variances of the estimators depend on the autocorrelation functions of the two components. Thus for moderately large values of N (e.g. N about 100) it is possible for two series, which are actually uncorrelated, to give rise to apparently 'large' cross-correlation coefficients which are actually spurious. Thus if a test is required for non-zero correlation between two time series, both series should first be filtered to convert them to white noise before computing the cross-correlation function (Jenkins and Watts, 1968, p. 340). For two uncorrelated series of white noise it can be shown that

$$E(r_{xy}(k)) \simeq 0$$
$$\text{Var}(r_{xy}(k)) \simeq 1/N$$

so that values outside the interval $\pm 2/\sqrt{N}$ are significantly different from zero.

The filtering procedure mentioned above is accomplished by treating each series separately and fitting an appropriate model. The new filtered series consists of the residuals from this model. For example suppose that one series appeared to be a first-order AR process with estimated parameters $\hat{\alpha}$. Then the filtered series is given by

$$x'_t = (x_t - \bar{x}) - \hat{\alpha}(x_{t-1} - \bar{x}).$$

8.1.3 Interpretation

The interpretation of the sample cross-correlation function can be fraught with danger unless one uses the prefiltering procedure described in Section 8.1.2. For example Coen *et al.* (1969) calculated cross-correlation functions between variables, such as (detrended) Financial Times (F.T.) share index and (detrended) U.K. car production and this resulted in a fairly smooth, roughly sinusoidal function with 'large' coefficients at lags 5 and 6. Coen *et al.* used this information to set up a regression model to 'explain' the variation in the F.T. share index. However Box and Newbold (1971) have shown that the 'large' cross-correlation coefficients are in fact quite spurious as the two series had not been properly filtered. By simply detrending the raw data, Coen *et al.* effectively assumed that the appropriate model for each series was of the form

$$x_t = \alpha + \beta t + a_t$$

where the a_t are *independent*. In fact the error structure was quite different.

If both series are properly filtered, we have seen that it is easy to test whether any of the cross-correlation coefficients are significantly different from zero. Following example 2 of Section 8.1.1, a peak in the estimated cross-correlation function at lag d may indicate that one series is related to the other when delayed by time d.

8.2 The cross-spectrum

The cross-correlation function is the natural tool for examining the relationship between two time series in the time domain. In this section we introduce a complementary function, called the cross spectral density function or cross-spectrum, which is the natural tool in the frequency domain.

By analogy with equation (6.11), we will define the cross-spectrum of a discrete bivariate process measured at unit intervals of time as the Fourier transform of the cross-covariance function, namely

$$f_{xy}(\omega) = \frac{1}{\pi}\left[\sum_{k=-\infty}^{\infty} \gamma_{xy}(k)e^{-i\omega k} \right] \qquad (8.6)$$

over the range $0 < \omega < \pi$. The cross-spectrum is sometimes called the cross-power spectrum, although its physical interpretation in terms of power is more difficult than for the autospectrum (see Lee, 1966, p. 348). Indeed a physical understanding of cross-spectra will probably not become clear until we have studied linear systems.

Note that $f_{xy}(\omega)$ is a *complex* function unlike the autospectrum which is real. This is because $\gamma_{xy}(k)$ is not an even function.

The reader must be told at this point that most authors define the cross-spectrum in the range $(-\pi, \pi)$ by analogy with equation (6.13) as

$$f_{xy}(\omega) = \frac{1}{2\pi}\left[\sum_{k=-\infty}^{\infty} \gamma_{xy}(k)e^{-i\omega k} \right] \qquad (8.7)$$

This definition has certain mathematical advantages, notably that it can handle complex-valued processes and that it has a simple inverse relationship of the form

$$\gamma_{xy}(k) = \int_{-\pi}^{\pi} e^{i\omega k} f_{xy}(\omega)d\omega \qquad (8.8)$$

whereas (8.6) does not have a simple inverse relationship. But (8.7) introduces negative frequencies, and for ease of understanding we shall use (8.6). As regards definition (8.7),

175

note that $f_{xy}(-\omega)$ is the complex conjugate of $f_{xy}(\omega)$ and so provides no extra information. Authors who use (8.7) only examine $f_{xy}(\omega)$ at positive frequencies. Note that Kendall and Stuart (1966) use a different definition to other authors by omitting the constant $1/\pi$ and transforming the cross-correlation function rather than the cross-covariance function.

We now describe several functions derived from the cross-spectrum which are helpful in interpreting the cross-spectrum. From (8.6), the *real* part of the cross-spectrum, called the *co-spectrum* is given by

$$c(\omega) = \frac{1}{\pi}\left[\sum_{k=-\infty}^{\infty} \gamma_{xy}(k) \cos \omega k\right] \tag{8.9}$$

$$= \frac{1}{\pi}\left\{\gamma_{xy}(0) + \sum_{k=1}^{\infty}\left[\gamma_{xy}(k) + \gamma_{yx}(k)\right] \cos \omega k\right\}.$$

Note that the formula given by Granger and Hatanaka (1964, p. 75) is incorrect.

The complex part of the cross-spectrum, with a minus sign, is called the *quadrature* spectrum and is given by

$$q(\omega) = \frac{1}{\pi}\left[\sum_{k=-\infty}^{\infty} \gamma_{xy}(k) \sin \omega k\right] \tag{8.10}$$

$$= \frac{1}{\pi}\left\{\sum_{k=1}^{\infty}\left[\gamma_{xy}(k) - \gamma_{yx}(k)\right] \sin \omega k\right\}$$

so that

$$f_{xy}(\omega) = c(\omega) - iq(\omega). \tag{8.11}$$

Note that Granger and Hatanaka (1964, p. 74) express the cross-spectrum as $c(\omega) + iq(\omega)$ in view of their alternative definition of the cross-covariance function.

An alternative way of expressing the cross-spectrum is in the form

$$f_{xy}(\omega) = \alpha_{xy}(\omega)e^{i\phi_{xy}(\omega)} \tag{8.12}$$

where $\alpha_{xy}(\omega)$ = cross-amplitude spectrum

$$= \sqrt{[c^2(\omega) + q^2(\omega)]} \qquad (8.13)$$

and $\quad \phi_{xy}(\omega)$ = phase spectrum

$$= \tan^{-1}[-q(\omega)/c(\omega)] \qquad (8.14)$$

From (8.14) it appears that $\phi_{xy}(\omega)$ is undetermined by a multiple of π. However if the cross-amplitude spectrum is required to be positive so that we take the positive square root in (8.13) then it can be seen that the phase is actually undetermined by a multitude of 2π using the equality of (8.11) and (8.12). This apparent non-uniqueness makes it difficult to graph the phase. However, when we consider linear systems in chapter 9, we will see that there are physical reasons why the phase *is* generally uniquely determined and not confined to the range $\pm\pi$ or $\pm\pi/2$. The phase is usually zero at $\omega = 0$ and is a continuous function as ω goes from 0 to π.

Another useful function derived from the cross-spectrum is the (squared) coherency which is given by

$$C(\omega) = [c^2(\omega) + q^2(\omega)]/[f_x(\omega)f_y(\omega)] \qquad (8.15)$$
$$= \alpha_{xy}^2(\omega)/f_x(\omega)f_y(\omega)$$

where $f_x(\omega)$, $f_y(\omega)$ are the power spectra of the individual processes, X_t and Y_t. It can be shown that

$$0 \leqslant C(\omega) \leqslant 1.$$

This quantity measures the linear correlation between the two components of the bivariate process at frequency ω and is analogous to the square of the usual correlation coefficient. The closer $C(\omega)$ is to one, the more closely related are the two processes at frequency ω.

Finally we will define a function called the gain spectrum which is given by

$$G_{xy}(\omega) = \sqrt{[f_y(\omega)C(\omega)/f_x(\omega)]}$$
$$= \alpha_{xy}(\omega)/f_x(\omega) \qquad (8.16)$$

which is essentially the regression coefficient of the process Y_t on the process X_t at frequency ω. A second gain function can also be defined by $G_{yx}(\omega) = \alpha_{xy}(\omega)/f_y(\omega)$ in which, using linear system terminology, Y_t is regarded as the input and X_t as the output. Note that Granger and Hatanaka (1964) reverse the order of the subscripts of $G_{xy}(\omega)$. Perhaps the notation of Parzen (1964) is clearer, using $G_{y|x}(\omega)$ for expression (8.16).

By this point, the reader will probably be rather confused by all the different functions which have been introduced in relation to the cross-spectrum. Whereas the cross-correlation function is a relatively straightforward development from the autocorrelation function, statisticians often find the cross-spectrum much harder to understand than the autospectrum. Usually *three* functions have to be plotted against frequency to describe the relationship between two series in the frequency domain. Sometimes the co-, quadrature and coherency spectra are most suitable. Sometimes the coherency, phase and cross-amplitude are more appropriate, while another possible trio is coherency, phase and gain. The physical interpretation of these functions will probably not become clear until we have studied linear systems in Chapter 9.

8.2.1 Examples
In this section, we derive the cross-spectrum and related functions for the two examples discussed in Section 8.1.1.

Example 1   Using (8.6) the cross-spectrum is given by

$$f_{xy}(\omega) = [0.5\sigma_Z^2 e^{-i\omega} + 0.5\sigma_Z^2 e^{-2i\omega}]/\pi.$$

Using (8.9), the co-spectrum is given by

$$c(\omega) = 0.5\sigma_Z^2[\cos \omega + \cos 2\omega]/\pi.$$

Using (8.10), the quadrature spectrum is given by

$$q(\omega) = 0.5\sigma_Z^2[\sin \omega + \sin 2\omega]/\pi.$$

Using (8.13), the cross-amplitude spectrum is given by

$$\alpha_{xy}(\omega) = \frac{0.5\sigma_Z^2}{\pi}\sqrt{[(\cos \omega + \cos 2\omega)^2 + (\sin \omega + \sin 2\omega)^2]}$$

which, after some algebra, gives

$$\alpha_{xy}(\omega) = \sigma_Z^2 \cos(\omega/2)/\pi.$$

Using (8.14), the phase spectrum is given by

$$\tan \phi_{xy}(\omega) = -(\sin \omega + \sin 2\omega)/(\cos \omega + \cos 2\omega).$$

In order to evaluate the coherency, we need to find the power spectra of the two processes. Since $X_t = Z_t$, it has a constant spectrum given by $f_x(\omega) = \sigma_Z^2/\pi$. The spectrum of Y_t is given by

$$f_y(\omega) = 0.5\sigma_Z^2(1 + \cos \omega)/\pi$$
$$= \sigma_Z^2 \cos^2(\omega/2)/\pi.$$

Thus, using (8.15), the coherency spectrum is given by

$$C(\omega) = 1 \qquad \text{for all } \omega \text{ in } (0, \pi).$$

This latter result may at first sight appear surprising. But both X_t and Y_t are generated from the *same* noise process and this explains why there is perfect correlation between the components of the two processes at any given frequency. Finally, using (8.16), the gain spectrum is given by

$$G_{xy}(\omega) = \cos(\omega/2).$$

Since the coherency is unity, $G_{xy}(\omega) = \sqrt{[f_y(\omega)/f_x(\omega)]}$.

Example 2 In this case we find

$$f_{xy}(\omega) = \sigma_Z^2 e^{-i\omega d}/\pi$$
$$c(\omega) = \sigma_Z^2 \cos \omega d/\pi$$
$$q(\omega) = \sigma_Z^2 \sin \omega d/\pi$$
$$\alpha_{xy}(\omega) = \sigma_Z^2/\pi$$
$$\tan \phi_{xy}(\omega) = -\tan \omega d. \qquad (8.17)$$

179

Then, as the two autospectra are given by

$$f_x(\omega) = \sigma_Z^2/\pi$$

and

$$f_y(\omega) = 2\sigma_Z^2/\pi$$

we find

$$C(\omega) = 1/2.$$

The function of particular interest in this example is the phase, which from (8.17), is a straight line with slope $(-d)$ when $\phi_{xy}(\omega)$ is unconstrained and is plotted against ω as a continuous function starting with zero phase at zero frequency (see Fig. 8.1(a)). If, however, the phase is constrained to lie within the interval $(-\pi, \pi)$ then a graph like Fig. 8.1(b) will result where the slope of each line is $(-d)$.

This result is often used in identifying relationships between time series. If the estimated phase approximates a straight line through the origin then this indicates a delay between the two series equal to the slope of the line. An example of this in practice is given by Barksdale and Guffey (1972). More generally, the time delay between two recording sites will change with frequency, due, for example, to varying speeds of propagation. This is called the *dispersive* case and real-life examples are discussed by Haubrich and Mackenzie (1965) and Hamon and Hannan (1974).

8.2.2 Estimation
As in Section 7.4, there are two basic approaches to estimating the cross-spectrum. Firstly we can take a Fourier transform of the truncated sample cross-covariance function (or of the cross-correlation function to get a normalized cross-spectrum). The estimated co-spectrum is then given by

$$\hat{c}(\omega) = \frac{1}{\pi} \left[\sum_{k=-M}^{M} \lambda_k c_{xy}(k) \cos \omega k \right] \qquad (8.18)$$

180

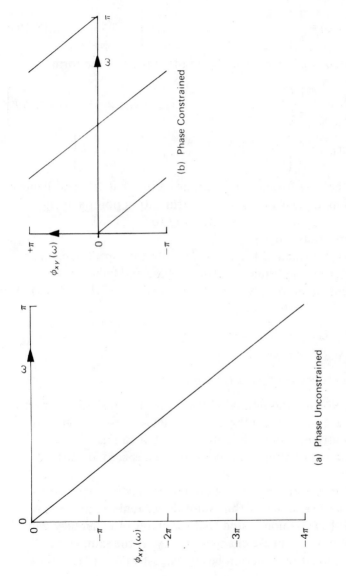

Figure 8.1   The phase spectrum for example 2 with d = 4, with (a) phase unconstrained; (b) phase constrained.

181

where M is the truncation point, and $\{\lambda_k\}$ is the lag window. The estimated quadrature spectrum is given by

$$\hat{q}(\omega) = \frac{1}{\pi} \left[\sum_{k=-M}^{M} \lambda_k c_{xy}(k) \sin \omega k \right] \qquad (8.19)$$

Equations (8.18) and (8.19) are often used in the form

$$\hat{c}(\omega) = \frac{1}{\pi} \left\{ \lambda_0 c_{xy}(0) + \sum_{k=1}^{M} \lambda_k [c_{xy}(k) + c_{xy}(-k)] \cos \omega k \right\}$$

$$\hat{q}(\omega) = \frac{1}{\pi} \left\{ \sum_{k=1}^{M} \lambda_k [c_{xy}(k) - c_{xy}(-k)] \sin \omega k \right\}$$

Note that the formula for $\hat{c}(\omega)$ given by Chatfield and Pepper (1970, p. 231) is incorrect. The truncation point and lag window are chosen in a similar way to that used in auto-spectral analysis.

Having estimated the co- and quadrature spectra, estimates of the cross-amplitude spectrum, phase and coherency following in an obvious way from equations (8.13), (8.14) and (8.15). We have

$$\hat{\alpha}_{xy}(\omega) = \sqrt{[\hat{c}^2(\omega) + \hat{q}^2(\omega)]}$$
$$\tan \hat{\phi}_{xy}(\omega) = -\hat{q}(\omega)/\hat{c}(\omega)$$
$$\text{and } \hat{C}(\omega) = \hat{\alpha}_{xy}^2(\omega)/\hat{f}_x(\omega)\hat{f}_y(\omega).$$

When plotting the estimated phase spectrum, similar remarks apply as to the (theoretical) phase. Phase estimates are apparently not uniquely determined but can usually be plotted as a continuous function which is zero at zero frequency.

Before estimating the coherency, it may be advisable to *align* the two series. If this is not done, Jenkins and Watts (1968) have demonstrated that estimates of coherency will be biased if the phase changes rapidly. If the sample cross-correlation function has its largest value at lag s, then

182

the two series are aligned by translating one of the series a distance s so that the peak in the cross-correlation function of the aligned series is at zero lag.

The second approach to cross-spectral analysis is to smooth a function called the *cross-periodogram*. The univariate periodogram of a series (x_t) can be written in the form

$$I(\omega_p) = (\Sigma x_t \, e^{i\omega_p t})(\Sigma x_t \, e^{-i\omega_p t})/N\pi \qquad (8.20)$$
$$= N(a_p^2 + b_p^2)/4\pi$$

using (7.17) and (7.10). By analogy with (8.20) we may define the cross-periodogram of two series (x_t) and (y_t) as

$$I_{xy}(\omega_p) = (\Sigma x_t \, e^{i\omega_p t})(\Sigma y_t \, e^{-i\omega_p t})/N\pi \qquad (8.21)$$

We then find that the real and imaginary parts of $I_{xy}(\omega_p)$ are given by

$$N(a_{px}a_{py} + b_{px}b_{py})/4\pi \text{ and } N(a_{px}b_{py} - a_{py}b_{px})/4\pi$$

where (a_{px}, b_{px}), (a_{py}, b_{py}) are the Fourier coefficients of $\{x_t\}$, $\{y_t\}$ at ω_p. These real and imaginary parts may then be smoothed to get consistent estimates of the co- and quadrature spectral density functions by

$$\hat{c}(\omega_p) = N \sum_{q=p-m^*}^{p+m^*} (a_{qx}a_{qy} + b_{qx}b_{qy})/4\pi m$$

$$\hat{q}(\omega_p) = N \sum_{q=p-m^*}^{p+m^*} (a_{qx}b_{qy} - a_{qy}b_{qx})/4\pi m$$

where $m = 2m^* + 1$. These estimates may then be used to estimate the cross-amplitude spectrum, phase, etc. as before.

This type of approach has been rather neglected in the literature yet the computational advantages are clear. Once a periodogram analysis has been made of the two individual processes, nearly all the work has been done as the estimates of $c(\omega)$ and $q(\omega)$ only involve the Fourier coefficients of the two series. The disadvantage of the approach is that alignment is only possible if the cross-correlation function is calculated separately. This can be done directly or by the use of two (fast) Fourier transforms by an analogous procedure to that described in Section 7.4.5. Tick (1967), suggests several smoothing procedures as alternatives to alignment. Further work needs to be done on this approach, but it seems likely to displace the traditional method for long series.

We will not say much about the properties of cross-spectral estimators. Asymptotic results are given by Jenkins and Watts (1968) while Granger and Hughes (1968) have carried out a simulation study on some short series. The following points are worth noting. Estimates of phase and cross-amplitude are imprecise when the coherency is relatively small. Estimates of coherency are constrained to lie between 0 and 1, and there may be a bias towards ½ which may be serious with short series. Rapid changes in phase may bias coherency estimates. Generally speaking, this is an area of time-series analysis where much work remains to be done.

8.2.3 Interpretation

Cross-spectral analysis is a technique for examining the relationship between two series over a range of frequencies. The technique may be used for two time series which 'arise on a similar footing' and then the coherency spectrum is perhaps the most useful function. It measures the linear correlation between two series at each frequency and is analogous to the square of the ordinary product−moment correlation coefficient.

The other functions introduced in this chapter, such as the phase spectrum, are more readily understood in the context of linear systems which will be discussed in Chapter 9. We will therefore defer further discussion of how to interpret cross-spectral estimates until Section 9.3.

BIVARIATE PROCESSES

Exercises

1. Show that the cross-covariance function of the discrete bivariate process (X_t, Y_t) where

$$X_t = Z_{1,t} + \beta_{11} Z_{1,t-1} + \beta_{12} Z_{2,t-1}$$
$$Y_t = Z_{2,t} + \beta_{21} Z_{1,t-1} + \beta_{22} Z_{2,t-1}$$

and $(Z_{1,t})$, $(Z_{2,t})$ are independent purely random processes with zero mean and variance σ_Z^2, is given by

$$
\gamma_{xy}(k) = \begin{cases}
\sigma_Z^2 (\beta_{11}\beta_{21} + \beta_{12}\beta_{22}) & k = 0 \\
\beta_{21} \sigma_Z^2 & k = 1 \\
\beta_{12} \sigma_Z^2 & k = -1 \\
0 & \text{otherwise.}
\end{cases}
$$

Hence evaluate the cross-spectrum.

2. Define the cross-correlation function, $\rho_{xy}(\tau)$, of a bivariate stationary process and show that $|\rho_{xy}(\tau)| \leqslant 1$ for all τ. Two MA processes

$$X_t = Z_t + 0.4 Z_{t-1}$$
$$Y_t = Z_t - 0.4 Z_{t-1}$$

are formed from a purely random process, $\{Z_t\}$, which has mean zero and variance σ_Z^2. Find the cross-covariance and cross-correlation functions of the bivariate process $\{X_t, Y_t\}$ and hence show that the cross-spectrum is given by

$$f_{xy}(\omega) = \sigma_Z^2 (0.84 + 0.8\, i \sin \omega)/\pi \quad (0 < \omega < \pi).$$

Evaluate the co-, quadrature, cross-amplitude, phase and coherency spectra.

CHAPTER 9

Linear Systems

9.1 Introduction

Many physical or mathematical systems may be thought of as devices which convert an input series into an output series. For example, the yield from a chemical reactor depends on the temperature of the process. When measurements are taken at discrete time intervals, we will denote the input and output by x_t, y_t respectively. When measurements are made continuously, we will use the notation $x(t)$ and $y(t)$.

An important class of systems are those where the output is a *linear* function of the input. The study of linear systems is an important branch of engineering (e.g. Schwarz and Friedland, 1965) and we shall only give a brief introduction. Of course, few physical systems are exactly linear, but many can be adequately approximated by a linear model over the range of interest.

We shall see that the study of linear systems is useful for examining the relationship between two time series, and for considering the properties of filtering procedures, such as detrending.

We shall only consider systems having one input and one output, though the extension to several inputs and outputs is fairly straightforward.

Let us first give a more precise definition of linearity. Suppose that $y_1(t)$, $y_2(t)$ are the outputs from a system corresponding to the inputs $x_1(t)$, $x_2(t)$. Then the system is said to be *linear* if, and only if, a linear combination of the inputs, $\lambda_1 x_1(t) + \lambda_2 x_2(t)$, produces the same linear

186

combination of the outputs, namely $\lambda_1 y_1(t) + \lambda_2 y_2(t)$, where λ_1, λ_2 are any constants.

We shall only consider linear systems which also have a property called *time-invariance*, which is defined as follows. Suppose that the input $x(t)$ to a system produces output $y(t)$. Then the system is said to be time-invariant if a delay of time τ in the input produces the same delay in the output. In other words input $x(t - \tau)$ produces output $y(t - \tau)$, so that the input–output relation does not change with time.

A system is said to be *physically realizable* or causal if the output depends only on present and past values of the input. A system is said to be *resonant* if it converts an input consisting of white noise into a series which has a spectrum with one or more high narrow peaks.

9.2 Linear systems in the time domain

The general form of a time-invariant linear system is given by

$$y(t) = \int_{-\infty}^{\infty} h(u)x(t - u)\mathrm{d}u \qquad (9.1)$$

in continuous time, or

$$y(t) = \sum_{k=-\infty}^{\infty} h_k x_{t-k} \qquad (9.2)$$

in discrete time. The weight function, $h(u)$ in continuous time or $\{h_k\}$ in discrete time, provides a description of the system in the time domain. This function is called the *impulse response function* of the system for reasons which will become apparent later. Note that some engineers prefer to call the discrete function, $\{h_k\}$, the *unit sample response function*.

It is clear that equations (9.1) and (9.2) are linear. The property of time-invariance ensures that the impulse response function does not depend on t. The system is physically realizable if

$$h(u) = 0 \qquad (u < 0)$$

or

$$h_k = 0 \qquad (k < 0).$$

Engineers have been principally concerned with continuous-time systems but are increasingly studying sampled-data control problems. Statisticians generally work with discrete data and so the subsequent discussion is mainly concerned with the discrete case.

We will only consider *stable* systems for which a bounded input produces a bounded output, although control engineers are frequently concerned with the control of unstable systems. A sufficient condition for stability is that the impulse response function should satisfy

$$\sum_k h_k < C$$

where C is a finite constant.

9.2.1 Some examples

The linear filters introduced in Section 2.5.2 are examples of linear systems. For example the simple moving average given by

$$y_t = (x_{t-1} + x_t + x_{t+1})/3$$

has impulse response function

$$h_k = \begin{cases} \dfrac{1}{3} & k = -1, 0, +1 \\ 0 & \text{otherwise} \end{cases}$$

Note that this filter is not 'physically realizable', although it can of course be used as a mathematical smoothing device.

Another general class of linear systems are those expressed as linear differential equations with constant coefficients in continuous time. For example, if T is a constant, then

$$T \frac{dy(t)}{dt} + y(t) = x(t).$$

is a description of a linear system. In discrete time, the analogue of differential equations are *difference* equations given by

$$y_t + \alpha_1 \nabla y_t + \alpha_2 \nabla^2 y_t + \ldots$$
$$= \beta_0 x_t + \beta_1 \nabla x_t + \beta_2 \nabla^2 x_t + \ldots \tag{9.3}$$

where $\nabla y_t = y_t - y_{t-1}$. Equation (9.3) can be rewritten as

$$y_t = a_1 y_{t-1} + a_2 y_{t-2} + \ldots$$
$$+ b_0 x_t + b_1 x_{t-1} + \ldots \tag{9.4}$$

It is clear that equation (9.4) can be rewritten in the form (9.2) by successive substitution. For example if

$$y_t = \tfrac{1}{2} y_{t-1} + x_t$$

then

$$y_t = x_t + \tfrac{1}{2} x_{t-1} + \tfrac{1}{4} x_{t-2} \ldots$$

so that the impulse response function is given by

$$h_k = \begin{cases} (\tfrac{1}{2})^k & k = 0, 1, \ldots \\ 0 & k < 0 \end{cases}$$

Two very simple linear systems are given by

$$y_t = x_{t-d} \tag{9.5}$$

called *simple delay*, where the integer d denotes the delay time time, and

$$y_t = g x_t \tag{9.6}$$

called *simple gain*, where g is a constant called the gain. The impulse response functions of (9.5) and (9.6) are

$$h_k = \begin{cases} 1 & k = d \\ 0 & \text{otherwise} \end{cases}$$

and

$$h_k = \begin{cases} g & k = 0 \\ 0 & \text{otherwise} \end{cases}$$

respectively.

In continuous time, the impulse response functions of simple delay and simple gain, namely

$$y(t) = x(t - \tau)$$

and

$$y(t) = gx(t)$$

can only be represented in terms of the Dirac delta function (see Appendix II). The functions are $\delta(u - \tau)$ and $g\delta(u)$ respectively.

An important class of impulse response functions, which often provides a reasonable approximation to physically realizable systems, is given by

$$h(u) = \begin{cases} [ge^{-(u-\tau)/T}]/T & (u > \tau) \\ 0 & (u < \tau) \end{cases}$$

A function of this type is called a *delayed exponential*, and depends on three constants, g, T, and τ. The constant τ is called the *delay*. When $\tau = 0$, we have simple exponential response. The constant g is called the *gain*, and represents the eventual change in output when a step change of unit size is made to the input. The constant T governs the rate at which the output changes. Fig. (9.1) shows how the output to a delayed exponential system changes when a step change of unity is made to the input.

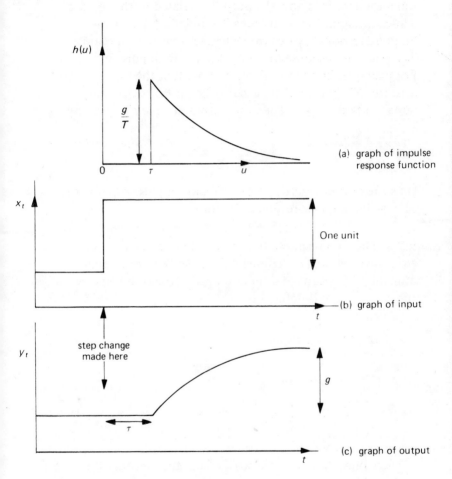

Figure 9.1　A delayed exponential response to a unit step change in input.

9.2.2 The impulse response function   The impulse response
function describes how the output is related to the input of
a linear system. (See equations 9.1 and 9.2.) The name
'impulse response' arises from the fact that the function
describes the response of the system to an impulse input.
For example, in discrete time, suppose that the input, x_t, is
zero for all t except at time zero when it takes the value
unity, so that $x_0 = 1$. Then the output at time t is given by

$$y_t = \Sigma h_k x_{t-k}$$
$$= h_t$$

Thus the output resulting from the unit impulse input is the
same as the impulse response function.

9.2.3 The step response function   An alternative,
equivalent, way of describing a linear system in the time
domain is by means of a function called the step response
function, which is defined by

$$S(t) = \int_{-\infty}^{t} h(u)du \tag{9.7}$$

in continuous time, and

$$S_t = \sum_{k \leqslant t} h_k \tag{9.8}$$

in discrete time.

The name 'step response' arises from the fact that the
function describes the response of the system to a unit step
change in the input. For example, in discrete time, suppose
that the input is given by

$$x_t = \begin{cases} 0 & t < 0 \\ 1 & t \geqslant 0 \end{cases}$$

192

then $y_t = \sum\limits_{k} h_k x_{t-k}$

$= \sum\limits_{k \leqslant t} h_k$

$= S_t,$

so that the output is equal to the step response function.

Engineers sometimes use this relationship to measure the properties of a physically realizable system. The input is held steady for some time and then a unit step change is made to the input. The output is then observed and this provides an estimate of the step response function, and hence, of its derivative, the impulse response function. A step change in the input may be easier to provide than an impulse.

The step response function for a delayed exponential system is given by

$$S(t) = g[1 - e^{-(t-\tau)/T}] \quad (t > \tau) \tag{9.9}$$

and the graph of y_t in Fig. (9.1) is also a graph of $S(t)$.

9.3 Linear systems in the frequency domain

9.3.1 The frequency response function   An alternative way of describing a linear system is by means of a function, called the *frequency response function* or *transfer function*, which is the Fourier transform of the impulse response function. It is defined by

$$H(\omega) = \int_{-\infty}^{\infty} h(u)e^{-i\omega u} du \quad (0 < \omega < \infty) \tag{9.10}$$

in continuous time, and

$$H(\omega) = \sum\limits_{k} h_k e^{-i\omega k} \quad (0 < \omega < \pi) \tag{9.11}$$

in discrete time.

193

The frequency response and impulse response functions are equivalent ways of describing a linear system, in a somewhat similar way that the autocovariance and spectral density functions are equivalent ways of describing a stationary stochastic process; one function being the Fourier transform of the other. We shall see that, for some purposes, $H(\omega)$ is much more useful than $h(u)$. First we prove the following theorem.

THEOREM 9.1   A sinusoidal input to a linear system gives rise, in the steady-state, to a sinusoidal output at the *same* frequency. The amplitude of the sinusoid may change and there may also be a phase shift.

Proof   The proof is given for continuous time, the extension to discrete time being straightforward. Suppose that the input to a linear system, with impulse response function, $h(u)$, is given by

$x(t) = \cos \omega t,$ for all t.

Then the output is given by

$$y(t) = \int_{-\infty}^{\infty} h(u) \cos \omega(t - u) \, du \qquad (9.12)$$

Now $\cos(A - B) = \cos A \cos B + \sin A \sin B$, so we may rewrite (9.12) as

$$y(t) = \cos \omega t \int_{-\infty}^{\infty} h(u) \cos \omega u \, du.$$

$$+ \sin \omega t \int_{-\infty}^{\infty} h(u) \sin \omega u \, du.$$

As the two integrals do not depend on t, it is now obvious that $y(t)$ is a mixture of sine and cosine terms at frequency ωt. Thus the output is a sinusoidal perturbation at the same frequency, ω, as the input.

194

If we write

$$A(\omega) = \int_{-\infty}^{\infty} h(u) \cos \omega u \, du$$

$$B(\omega) = \int_{-\infty}^{\infty} h(u) \sin \omega u \, du$$

$$G(\omega) = \sqrt{[A^2(\omega) + B^2(\omega)]} \qquad (9.13)$$

and $\tan \phi(\omega) = -B(\omega)/A(\omega)$ \qquad (9.14)

then

$$y(t) = A(\omega) \cos \omega t + B(\omega) \sin \omega t$$
$$= G(\omega) \cos(\omega t + \phi(\omega)) \qquad (9.15)$$

Equation (9.15) shows that a cosine wave is amplified by a factor $G(\omega)$, which is called the *gain* of the system. The equation also shows that the cosine wave is shifted by an angle $\phi(\omega)$, which is called the *phase shift*. Note that both the gain and phase shift may vary with frequency. From equation (9.14) we see that the phase shift is apparently not uniquely determined. If we take the positive square root in equation (9.13), so that the gain is required to be positive, then the phase shift is undetermined by a multiple of 2π (see also Sections 8.2 and 9.3.2).

We have so far considered an input cosine wave. By a similar argument it can be shown that an input sine wave, $x(t) = \sin \omega t$, gives an output $y(t) = G(\omega) \sin(\omega t + \phi(\omega))$, so that there is the same gain and phase shift. More generally if we consider an input complex exponential given by

$$x(t) = e^{i\omega t}$$
$$= \cos \omega t + i \sin \omega t$$

then the output is given by

$$y(t) = G(\omega)\{\cos[\omega t + \phi(\omega)] + i \sin[\omega t + \phi(\omega)]\}$$
$$= G(\omega)e^{i[\omega t + \phi(\omega)]}$$
$$= G(\omega)e^{i\phi(\omega)}x(t). \qquad (9.16)$$

Now

$$G(\omega)e^{i\phi(\omega)} = A(\omega) - iB(\omega)$$

$$= \int_{-\infty}^{\infty} h(u)(\cos \omega u - i \sin \omega u)du$$

$$= \int_{-\infty}^{\infty} h(u)e^{-i\omega u}\,du$$

$$= H(\omega) \qquad\qquad (9.17)$$

So when the input in equation (9.16) is of the form $e^{i\omega t}$, the output is given simply by (frequency response function x input) and we have (in the steady-state situation)

$$y(t) = H(\omega)\,x(t) \qquad\qquad (9.18)$$

This completes the proof of Theorem 9.1.

Transients   The reader should note that theorem 9.1 only applies in the *steady-state* where it is assumed that the input sinusoid was applied at $t = -\infty$. If in fact the sinusoid is applied at say $t = 0$, then the output will take some time to settle to the steady-state form given by the theorem. The difference between the observed output and the steady-state output is called the *transient* component. The system is stable if this transient component tends to zero as $t \rightarrow \infty$. If the relationship between input and output is expressed as a differential (or difference) equation, then the steady-state solution corresponds to the particular integral, while the transient component corresponds to the complementary function.

It is easier to describe the transient behaviour of a linear system by using the *Laplace* transform of the impulse response function. Engineers also prefer the Laplace transform as it is defined for unstable systems. However statisticians have customarily dealt with steady-state behaviour and used Fourier transforms, and we will continue this custom. However this is certainly an aspect of linear systems which statisticians should look at more closely.

DISCUSSION OF THEOREM 9.1   Theorem 9.1 helps to introduce the importance of the frequency response function. For inputs consisting of an impulse or step change is it easy to calculate the output using the impulse response function. But for a sinusoidal input, it is much easier to calculate the output using the frequency response function. [Compare equations (9.12) and (9.18).] More generally for an input consisting of several sinusoidal perturbations, namely

$$x(t) = \sum_j A_j(\omega_j)e^{i\omega_j t}$$

it is easy to calculate the output using the frequency response function as

$$y(t) = \sum_j A_j(\omega_j)H(\omega_j)e^{i\omega_j t}.$$

Thus a complicated convolution in the time domain, as in equation (9.12), reduces to a simple multiplication in the frequency domain, and we shall see that linear systems are often easier to study in the frequency domain using the frequency response function.

Returning to the definition of the frequency response function as given by equations (9.10) and (9.11), note that some authors define $H(\omega)$ for negative as well as positive frequencies. But for real-valued processes we need only consider $H(\omega)$ for $\omega > 0$. Note that, in discrete time, $H(\omega)$ is only defined for frequencies up to the Nyquist frequency, π (or $\pi/\Delta t$ if there is an interval Δt between successive observations). We have already introduced the Nyquist frequency in Section 7.2.1. Applying similar ideas to a linear system, it is clear that a sinusoidal input which has a higher frequency than π will have a corresponding sinusoid at a frequency in $(0, \pi)$ which gives identical readings at unit

197

intervals of time and which will therefore give rise to an identical output.

We have already noted that $H(\omega)$ is sometimes called the frequency response function and sometimes the transfer function. We will use the former term as it is more descriptive, indicating that the function shows how a linear system responds to sinusoids at different frequencies. In any case the term transfer function is used by some authors in a different way. Engineers use the term to denote the *Laplace* transform of the impulse response function (see Appendix I). For a physically realizable stable system, the Fourier transform of the impulse response function may be regarded as a special case of the Laplace transform. A necessary and sufficient condition for a linear system to be stable is that the Laplace transform of the impulse response function should have no poles in the right half plane or on the imaginary axis. For an unstable system, the Fourier transform does not exist, but the Laplace transform does. But we will only consider stable systems, in which case the Fourier transform is adequate. Note that Jenkins and Watts (1968) use the term 'transfer function' to denote the Z-transform of the impulse response function (see Appendix I) in the discrete case. Z-transforms are also used by engineers for discrete-time systems (e.g. Schwarz and Friedland, 1965).

9.3.2 Gain and phase diagrams   The frequency response function, $H(\omega)$, of a linear system is a complex function given by

$$H(\omega) = G(\omega)e^{i\phi(\omega)}$$

where $G(\omega)$, $\phi(\omega)$ are the gain and phase respectively. In order to understand the frequency properties of the system, it is useful to plot $G(\omega)$ and $\phi(\omega)$ against ω to obtain what are called the *gain diagram* and the *phase diagram*. If $G(\omega)$ is 'large' for low values of ω, but 'small' for high values of ω, as in Fig. 9.2(a), then we have what is called a *low-pass* filter.

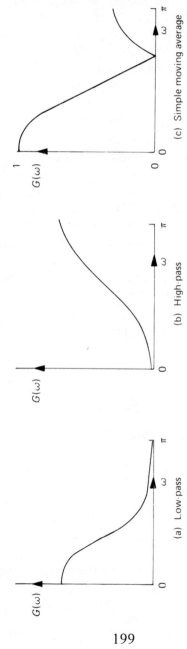

Figure 9.2 Gain diagrams for (a) a low-pass filter; (b) a high-pass filter; (c) a simple moving average of three successive observations.

This description is self-explanatory in that, if the input is a mixture of variation at several different frequencies, only those components with a low frequency will 'get through' the filter.

Conversely, if $G(\omega)$ is 'small' for low values of ω, but 'large' for high values of ω, then we have a *high-pass* filter as in Fig. 9.2(b).

Plotting the phase diagram is complicated by the fact that the phase in equation (9.14) is not uniquely determined. If the gain is always taken to be positive, then the phase is undetermined by a multiple of 2π and is often constrained to the range $(-\pi, \pi)$. The (complex) value of $H(\omega)$ is examined to see which quadrant it is in. If $G(\omega)$ can be positive or negative, the phase is undetermined by a multiple of π, and is often constrained to the range $(-\pi/2, \pi/2)$. These different conventions are not discussed adequately in many books (Hause, 1971, p. 214). In fact there are physical reasons why engineers prefer to plot the phase as a continuous unconstrained function allowing $G(\omega)$ to be positive or negative, and using the fact that $\phi(0) = 0$ provided $G(0)$ is finite.

9.3.3 Some examples

Example I   Consider the simple moving average

$$y_t = (x_{t-1} + x_t + x_{t+1})/3$$

which is a linear system with impulse response function

$$h_k = \begin{cases} \dfrac{1}{3} & k = -1, 0, +1 \\ 0 & \text{otherwise} \end{cases}$$

The frequency response function of this filter is (using equation 9.11)

$$H(\omega) = \frac{1}{3} e^{-i\omega} + \frac{1}{3} + \frac{1}{3} e^{i\omega}$$

$$= \frac{1}{3} + \frac{2}{3} \cos \omega \qquad (0 < \omega < \pi).$$

This function happens to be real, not complex, and so the phase appears to be given by

$$\phi(\omega) = 0 \qquad (0 < \omega < \pi).$$

However $H(\omega)$ is negative for $\omega > 2\pi/3$, and so if we adopt the convention that the gain should be positive, then we have

$$G(\omega) = \left| \frac{1}{3} + \frac{2}{3} \cos \omega \right|$$

$$= \begin{cases} \dfrac{1}{3} + \dfrac{2}{3} \cos \omega & 0 < \omega < 2\pi/3 \\[2mm] -\dfrac{1}{3} - \dfrac{2}{3} \cos \omega & 2\pi/3 < \omega < \pi \end{cases}$$

and

$$\phi(\omega) = \begin{cases} 0 & 0 < \omega < 2\pi/3 \\ \pi & 2\pi/3 < \omega < \pi. \end{cases}$$

The gain is plotted in Fig. 9.2(c) and is of low-pass type. This is to be expected as a moving average smooths out local fluctuations (high frequency variation) and measures the trend (the low frequency variation). In fact it is probably more sensible to allow the gain to go negative in $(2\pi/3, \pi)$ so that the phase is zero for all ω in $(0, \pi)$.

Example 2 A linear system showing simple exponential response has impulse response function

$$h(u) = g \, e^{-u/T}/T \qquad (u > 0).$$

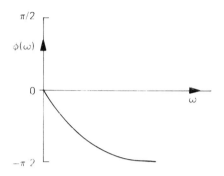

*Figure 9.3   Phase diagram for a
simple exponential response system.*

Using (9.10), the frequency response function is

$$H(\omega) = g(1 - i\omega T)/(1 + \omega^2 T^2) \qquad (\omega > 0).$$

Hence

$$G(\omega) = g/\sqrt{(1 + \omega^2 T^2)},$$

and

$$\tan \phi(\omega) = -T\omega.$$

As the frequency increases, $G(\omega)$ decreases so that the
system is of low-pass type. As regards the phase, if we take
$\phi(\omega)$ to be zero at zero frequency, then the phase becomes
increasingly negative as ω increases until the output is out of
phase with the input (see Fig. 9.3).

Example 3   Consider the linear system consisting of pure
delay, so that

$$y(t) = x(t - \tau), \text{ where } \tau \text{ is a constant.}$$

Jenkins and Watts (1968) give the impulse response function
as

$$h(u) = \delta(u - \tau)$$

where δ denotes the Dirac delta function (see Appendix II).

202

Then the frequency response function is given by

$$H(\omega) = \int_{-\infty}^{\infty} \delta(u - \tau)e^{-i\omega u}du$$

$$= e^{-i\omega\tau}.$$

In fact $H(\omega)$ can be derived without bringing in delta functions by using theorem 9.1. Suppose that input $x(t) = e^{i\omega t}$ is applied to the system. Then the output is $y(t) = e^{i\omega(t-\tau)} = e^{-i\omega\tau} \times$ input. Thus, by analogy with equation (9.18), we have $H(\omega) = e^{-i\omega\tau}$.

For this linear system, the gain is constant, namely

$$G(\omega) = 1,$$

while the phase is given by

$$\phi(\omega) = -\omega\tau.$$

9.3.4 General relation between input and output   We have, so far, considered only sinusoidal inputs in the frequency domain. In this section we consider any type of input and show that it is generally easier to work with linear systems in the frequency domain, than in the time domain.

The general relation between input and output in the time domain is given by equation (9.1), namely

$$y(t) = \int_{-\infty}^{\infty} h(u)x(t - u)du \tag{9.19}$$

When $x(t)$ is not of a simple form, this integral may be hard to evaluate. Now consider the Fourier transform of the output by

$$Y(\omega) = \int_{-\infty}^{\infty} y(t)e^{-i\omega t}dt$$

$$= \int_{-\infty}^{\infty}\int_{-\infty}^{\infty} h(u)x(t - u)e^{-i\omega t}du\, dt$$

$$= \int_{-\infty}^{\infty}\int_{-\infty}^{\infty} h(u)e^{-i\omega u}\, x(t - u)e^{-i\omega(t - u)}\, du\, dt.$$

But

$$\int_{-\infty}^{\infty} x(t-u)e^{-i\omega(t-u)}\,dt = \int_{-\infty}^{\infty} x(t)e^{-i\omega t}\,dt$$

for all values of u, and is therefore the Fourier transform of $x(t)$ which we will denote by $X(\omega)$. And

$$\int_{-\infty}^{\infty} h(u)e^{-i\omega u}\,du = H(\omega)$$

so that

$$Y(\omega) = H(\omega)X(\omega) \tag{9.20}$$

Thus the integral in (9.19) corresponds to a multiplication in the frequency domain provided that the Fourier transforms exist. A similar result holds in discrete time.

A more useful general relation between input and output, akin to equation (9.20), can be obtained when the input $x(t)$ is a stationary process with continuous power spectrum. This result will be given as theorem 9.2.

THEOREM 9.2   Consider a stable linear system with gain function, $G(\omega)$. Suppose that the input $X(t)$ is a stationary process with continuous power spectrum $f_x(\omega)$. Then the output $Y(t)$ is also a stationary process, whose power spectrum, $f_y(\omega)$, is given by

$$f_y(\omega) = G^2(\omega)f_x(\omega) \tag{9.21}$$

Proof   The proof will be given for continuous time, but the same result holds for discrete time. Let us denote the impulse response and frequency response functions of the system by $h(u)$, $H(\omega)$ respectively. Thus $G(\omega) = |H(\omega)|$.

It is easy to show that a stationary input to a stable linear system gives rise to a stationary output and this will not be shown here. For mathematical convenience, let us assume that the input has mean zero. Then the output

$$Y(t) = \int_{-\infty}^{\infty} h(u)X(t - u) \, du$$

also has mean zero.

Denote the autocovariance functions of $X(t)$, $Y(t)$ by $\gamma_x(\tau)$, $\gamma_y(\tau)$ respectively. Then

$$\gamma_y(\tau) = E[Y(t)Y(t + \tau)] \qquad \text{since} \quad E[Y(t)] = 0$$

$$= E\left\{ \int_{-\infty}^{\infty} h(u)X(t - u)du \int_{-\infty}^{\infty} h(u')X(t + \tau - u')du' \right\}$$

$$= \iint_{-\infty}^{\infty} h(u)h(u')E[X(t - u)X(t + \tau - u')] \, du \, du'$$

But

$$E[X(t - u)X(t + \tau - u')] = \gamma_x(\tau - u' + u).$$

Thus

$$\gamma_y(\tau) = \iint_{-\infty}^{\infty} h(u)h(u')\gamma_x(\tau - u' + u)du \, du' \qquad (9.22)$$

The relationship (9.22) between the autocovariance functions of input and output is not of a simple form. However if we take Fourier transforms of both sides of (9.22), by multiplying by $e^{-i\omega\tau}/\pi$ and integrating with respect to τ, we find for the left hand side

$$\frac{1}{\pi} \int_{-\infty}^{\infty} \gamma_y(\tau)e^{-i\omega\tau} \, d\tau = fy(\omega)$$

from equation (6.17), while for the right hand side

$$\iint_{-\infty}^{\infty} h(u)h(u')\left[\frac{1}{\pi} \int_{-\infty}^{\infty} \gamma_x(\tau - u' + u)e^{-i\omega\tau} \, d\tau \right] du \, du'$$

$$= \iint_{-\infty}^{\infty} h(u)e^{i\omega u} h(u')e^{-i\omega u'}$$

$$\times \left[\frac{1}{\pi} \int_{-\infty}^{\infty} \gamma_x(\tau - u' + u)e^{-i\omega(\tau - u' + u)} \, d\tau \right] du \, du'$$

But

$$\frac{1}{\pi} \int_{-\infty}^{\infty} \gamma_x(\tau - u' + u)e^{-i\omega(\tau - u' + u)}d\tau$$

$$= \frac{1}{\pi} \int_{-\infty}^{\infty} \gamma_x(\tau)e^{-i\omega\tau}d\tau = f_x(\omega)$$

for all u, u', and

$$\int_{-\infty}^{\infty} h(u)e^{i\omega u} = \overline{H(\omega)}$$

$$= G(\omega)e^{-i\phi(\omega)}$$

Thus

$$f_y(\omega) = \overline{H(\omega)}\, H(\omega) f_x(\omega)$$

$$= G^2(\omega) f_x(\omega)$$

This completes the proof.

The relationship between the spectra of input and output of a linear system is a very simple one. Once again a result in the frequency domain [equation (9.21)] is much simpler than the corresponding result in the time domain [equation (9.22)].

Theorem 9.2 can be used to evaluate the spectrum of some types of stationary process in a simpler way to that used in Chapter 6, where the method used was to evaluate the autocovariance function of the process and then find its Fourier transform. Several examples will now be given.

Moving average process   A MA process of order m is given by

$$X_t = \beta_0 Z_t + \ldots + \beta_m Z_{t-m}$$

where Z_t denotes a purely random process with variance σ_Z^2. This equation may be regarded as specifying a linear system with $\{Z_t\}$ as input and $\{X_t\}$ as output, whose frequency

206

response function is given by

$$H(\omega) = \sum_{j=0}^{m} \beta_j e^{-i\omega j}$$

As $\{Z_t\}$ is a purely random process, its spectrum is given by (see equation 6.19)

$$f_Z(\omega) = \sigma_Z^2 / \pi$$

Thus, using (9.21), the spectrum of $\{X_t\}$ is given by

$$f_x(\omega) = \left| \sum_{j=0}^{m} \beta_j e^{-i\omega j} \right|^2 \sigma_Z^2 / \pi$$

For example, for the first order MA process,

$$X_t = Z_t + \beta Z_{t-1}, \tag{9.22a}$$

we have

$$H(\omega) = 1 + \beta e^{-i\omega}$$

and

$$\begin{aligned} G^2(\omega) = |H(\omega)|^2 &= (1 + \beta \cos \omega)^2 + \beta^2 \sin^2 \omega \\ &= 1 + 2\beta \cos \omega + \beta^2 \end{aligned}$$

so that $f_x(\omega) = (1 + 2\beta \cos \omega + \beta^2)\sigma_Z^2/\pi$ as already derived in Section 6.5.

This type of approach can also be used when $\{Z_t\}$ is not a purely random process. For example, suppose that the $\{Z_t\}$ process in equation (9.22a) has arbitrary spectrum $f_Z(\omega)$. Then the spectrum of $\{X_t\}$ is given by

$$f_x(\omega) = (1 + 2\beta \cos \omega + \beta^2)f_Z(\omega).$$

Autoregressive process   The first order AR process

$$X_t = \alpha X_{t-1} + Z_t$$

may be regarded as a linear system producing output X_t from input Z_t. It may also be regarded as a linear system 'the other way round', producing output Z_t from input X_t by

$$Z_t = X_t - \alpha X_{t-1}$$

and this formulation which has frequency response function

$$H(\omega) = 1 - \alpha e^{-i\omega}$$

gives the desired result in the easiest mathematical way. Thus

$$G^2(\omega) = 1 - 2\alpha \cos \omega + \alpha^2$$

and so

$$f_Z(\omega) = (1 - 2\alpha \cos \omega + \alpha^2)f_x(\omega). \qquad (9.23)$$

But if $\{Z_t\}$ denotes a purely random process with spectrum $f_Z(\omega) = \sigma_Z^2/\pi$, then equation (9.23) may be rewritten to evaluate $f_x(\omega)$ as

$$f_x(\omega) = \sigma_Z^2/\pi(1 - 2\alpha \cos \omega + \alpha^2),$$

which has already been obtained as equation (6.23) by the earlier method.

This approach may also be used for higher order AR processes.

Differentiation   Consider the linear system which converts a continuous input $X(t)$ into output $Y(t)$ by

$$Y(t) = \frac{dX(t)}{dt} \qquad (9.24)$$

A differentiator is of considerable mathematical interest, although in practice only approximations to it are physically realizable.

If the input is sinusoidal, $X(t) = e^{i\omega t}$, then the output is given by

$$Y(t) = i\omega e^{i\omega t}$$

so that, using (9.18), the frequency response function is

$$H(\omega) = i\omega.$$

If the input is a stationary process, with spectrum $f_x(\omega)$, then it appears that the output has spectrum

$$f_y(\omega) = |i\omega|^2 f_x(\omega)$$
$$= \omega^2 f_x(\omega). \tag{9.25}$$

However this result assumes that the linear system (9.24) is stable, when in fact it is only stable for certain types of input process. For example, it is clear that the response to a unit step change is an unbounded impulse. In order for the system to be stable, the variance of the output must be finite. Now

$$\text{Var}[Y(t)] = \int_0^\infty f_y(\omega) \, d\omega$$

$$= \int_0^\infty \omega^2 f_x(\omega) \, d\omega.$$

But, using equation (6.18), we have

$$\gamma_x(k) = \int_0^\infty f_x(\omega) \cos \omega k \, d\omega$$

and

$$\frac{d^2 \gamma_x(k)}{dk^2} = -\int_0^\infty \omega^2 f_x(\omega) \cos \omega k \, d\omega$$

so that

$$\text{Var}[Y(t)] = -\left[\frac{d^2 \gamma_x(k)}{dk^2} \right]_{k=0}$$

Thus $Y(t)$ has finite variance provided that $\gamma_x(k)$ can be differentiated twice at $k = 0$, and only then does equation (9.25) hold.

9.3.5 Linear systems in series The advantages of working in the frequency domain are also evident when we consider two or more linear systems in series (or in cascade). For example Fig. 9.4 shows two linear systems in series, where the input $x(t)$ to system I produces output $y(t)$ which in turn is the input to system II producing output $z(t)$. It is often of interest to evaluate the properties of the overall system, which is also linear, where $x(t)$ is the input and $z(t)$ is the output. We will denote the impulse response and frequency response functions of systems I and II by $h_1(u)$, $h_2(u)$, $H_1(\omega)$, and $H_2(\omega)$.

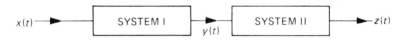

Figure 9.4   Two linear systems in series.

In the time domain, the relationship between $x(t)$ and $z(t)$ would be in the form of a double integral involving $h_1(u)$ and $h_2(u)$, which is rather complicated. But in the frequency domain we can denote the Fourier transforms of $x(t)$, $y(t)$, $z(t)$ by $X(\omega)$, $Y(\omega)$, $Z(\omega)$ and use equation (9.20). Then

$$Y(\omega) = H_1(\omega)X(\omega)$$

and

$$Z(\omega) = H_2(\omega)Y(\omega)$$
$$= H_2(\omega)H_1(\omega)X(\omega).$$

Thus it is easy to see that the overall frequency response function of the combined system is

$$H(\omega) = H_1(\omega)H_2(\omega). \tag{9.26}$$

If

$$H_1(\omega) = G_1(\omega)e^{i\phi_1(\omega)}$$

210

and

$$H_2(\omega) = G_2(\omega)e^{i\phi_2(\omega)}$$

then

$$H(\omega) = G_1(\omega)G_2(\omega)e^{i[\phi_1(\omega) + \phi_2(\omega)]}$$

Thus the overall gain is the *product* of the component gains, while the overall phase is the *sum* of the component phases.

The above results are easily extended to the situation where there are k linear systems in series with respective frequency response functions, $H_1(\omega), \ldots, H_k(\omega)$. The overall frequency response function is

$$H(\omega) = H_1(\omega)H_2(\omega) \ldots \times H_k(\omega).$$

9.3.6 Design of filters   The results of this section enable us to consider in more depth the properties of the filters introduced in Section 2.5.2. Given a time series, $\{x_t\}$, the filters for estimating or removing trend are of the form

$$y_t = \sum_k h_k x_{t-k}$$

and are clearly linear systems with frequency response function

$$H(\omega) = \sum_k h_k e^{-i\omega k}.$$

If the time-series has spectrum $f_x(\omega)$, then the spectrum of the smoothed series is given by

$$f_y(\omega) = G^2(\omega)f_x(\omega) \tag{9.27}$$

where $G(\omega) = |H(\omega)|$.

How do we set about choosing an appropriate filter for a time-series? The design of a filter involves a choice of $\{h_k\}$ and hence of $H(\omega)$ and $G(\omega)$. Two types of 'ideal' filter are shown in Fig. 9.5. Both have sharp cut-offs, the low-pass filter completely eliminating high frequency variation and the

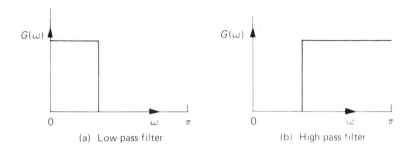

Figure 9.5   Two types of ideal filter; (a) a low-pass filter or trend estimator; (b) a high-pass filter or trend eliminator.

high-pass filter completely eliminating low frequency variation.

But 'ideal' filters of this type are impossible to achieve with a finite set of weights. Instead the smaller the number of weights used, the less sharp will generally be the cut-off property of the filter. For example the gain diagram of a simple moving average of three successive observations (Fig. 9.2(c)) is of low-pass type but has a much less sharp cut-off than the 'ideal' low-pass filter (Fig. 9.5(a)). More sophisticated moving averages such as Spencer's 15 point moving average have much better cut-off properties.

The differencing filter (Section 2.5.3), for removing a trend, of the form

$$y_t = x_t - x_{t-1}$$

has frequency response function

$$H(\omega) = 1 - e^{-i\omega}$$

and gain function

$$G(\omega) = \sqrt{[2(1 - \cos \omega)]}$$

which is plotted in Fig. 9.6. This is indeed of high-pass type, but the cut-off property is rather poor and this should be borne in mind when working with first differences.

212

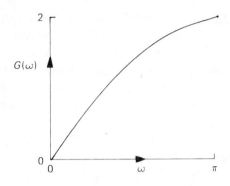

Figure 9.6   The gain diagram for the
difference operator.

9.4 The identification of linear systems

We have so far assumed that the structure of the linear
system under consideration is known. Given the impulse
response function of a system, or equivalently the frequency
response function, we can find the output corresponding to a
given input. In particular, when considering the properties of
filters for estimating trend and seasonality, a formula for the
'system' is given.

But many problems concerning linear systems are of a
completely different type. The structure of the system is *not*
known and the problem is to examine the relationship
between input and output so as to infer the properties of the
system. This procedure is called the *identification* of the
system. For example suppose we are interested in the effect
of temperature on the yield from a chemical process. Here we
have a physical system which we assume, initially at least, is
approximately linear over the range of interest. By examining
the relationship between observations on temperature (the
input) and yield (the output) we can infer the properties of
the chemical process.

The identification process is straightforward if the input to
the system can be controlled and if the system is 'not

213

contaminated by noise'. In this case, we can simply apply an impulse or step change input, observe the output, and hence estimate the impulse response or step response function. Alternatively we can apply sinusoidal inputs at different frequencies and observe the amplitude and phase shift of the corresponding sinusoidal outputs. This enables us to evaluate the gain and phase diagrams.

But many systems are contaminated by noise as illustrated in Fig. 9.7, where $N(t)$ denotes a noise process. This noise process may not be white noise (i.e. may not be a purely random process), but is usually assumed to be uncorrelated with the input process, $X(t)$.

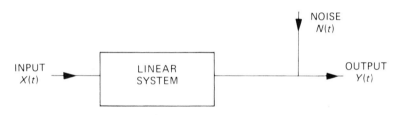

Figure 9.7   A linear system with added noise.

A further difficulty arises when the input is observable but not controllable. In other words one cannot make changes, such as a step change, to the input. For example attempts have been made to treat the economy as a linear system and to examine the relationship between variables like price increases (input) and wage increases (output). But price increases can only be controlled to a certain extent by governmental decisions, and there is also a feedback problem in that wage increases may in turn affect price increases (see Section 9.4.3).

When the system is affected by noise and/or the input is not controllable, more refined techniques are required to identify the system. We will describe two alternative approaches, one in the frequency domain and one in the time

214

domain. In Section 9.4.1, we show how cross-spectral analysis of input and output may be used to estimate the frequency response function of a linear system. In Section 9.4.2, we describe a method proposed by Box and Jenkins (1970) for estimating the impulse response function of a linear system.

9.4.1 Estimating the frequency response function

Suppose that we have a linear system with added noise, as depicted in Fig. 9.7, where the noise is assumed to be uncorrelated with the input and to have mean zero. Suppose also that we have observations on input and output over some time-period and wish to estimate the frequency response function of the system. We will denote the (unknown) impulse response and frequency response functions of the system by $h(u)$, $H(\omega)$ respectively.

The reader may think that equation (9.21), namely

$$f_y(\omega) = G^2(\omega)f_x(\omega)$$

can be used to estimate the gain of the system. But this equation does not hold in the presence of noise, $N(t)$, and does not in any case give information about the phase of the system. Instead we will derive a relationship involving the cross-spectrum of input and output.

In continuous time, the output $Y(t)$ is given by

$$Y(t) = \int_0^\infty h(u)X(t-u)du + N(t) \tag{9.28}$$

Note that we are only considering physically realizable systems so that $h(u)$ is zero for $u < 0$. For mathematical convenience we assume $E[X(t)] = 0$ so that $E[Y(t)] = 0$, but the following results also hold if $E[X(t)] \neq 0$. Multiplying through equation (9.28) by $X(t - \tau)$ and taking expectations, we have

$$E[N(t)X(t - \tau)] = 0$$

215

since $N(t)$ is assumed to be uncorrelated with input, so that

$$\gamma_{xy}(\tau) = \int_0^\infty h(u)\,\gamma_{xx}(\tau - u)\mathrm{d}u \qquad (9.29)$$

where γ_{xy} is the cross-covariance function of $X(t)$ and $Y(t)$ and γ_{xx} is the autocovariance function of $X(t)$. Equation (9.29) is called the Wiener–Hopf integral equation and, given γ_{xy} and γ_{xx}, can in principle be solved to find the impulse response function, $h(u)$. But it is often easier to work with the corresponding relationship in the frequency domain.

First we revert to discrete time and note that the discrete-time analogue of equation (9.29) is

$$\gamma_{xy}(\tau) = \sum_{k=0}^\infty h_k \gamma_{xx}(\tau - k) \qquad (9.30)$$

Take Fourier transforms of both sides of this equation by multiplying by $e^{-i\omega\tau}/\pi$ and summing from $\tau = -\infty$ to $+\infty$. Then we find

$$f_{xy}(\omega) = \sum_{\tau=-\infty}^\infty \sum_{k=0}^\infty h_k e^{-i\omega k}\,\gamma_{xx}(\tau - k)e^{-i\omega(\tau - k)}/\pi$$

$$= \sum_{k=0}^\infty h_k e^{-i\omega k} f_x(\omega)$$

$$= H(\omega)f_x(\omega) \qquad (9.31)$$

where f_{xy} is the cross-spectrum of input and output and f_x is the (auto)spectrum of the input. Thus, once again, a convolution in the time domain corresponds to a multiplication in the frequency domain.

Estimates of $f_{xy}(\omega)$ and $f_x(\omega)$ can now be used to estimate $H(\omega)$ using (9.31). Denote the estimated spectrum of the input by $\hat{f}_x(\omega)$, and the estimate of f_{xy}, obtained by cross-spectral analysis, by $\hat{f}_{xy}(\omega)$. Then

$$\hat{H}(\omega) = \hat{f}_{xy}(\omega)/\hat{f}_x(\omega).$$

We usually write

$$H(\omega) = G(\omega)e^{i\phi(\omega)}$$

and estimate the gain and phase separately. We have

$$\hat{G}(\omega) = |\hat{H}(\omega)| = |\hat{f}_{xy}(\omega)/\hat{f}_x(\omega)|$$
$$= |\hat{f}_{xy}(\omega)|/\hat{f}_x(\omega) \qquad \text{since } f_x(\omega) \text{ is real}$$
$$= \hat{\alpha}_{xy}(\omega)/\hat{f}_x(\omega) \qquad\qquad (9.32)$$

where $\alpha_{xy}(\omega)$ is the cross-amplitude spectrum [see equation (8.13)].

We also find

$$\tan\hat{\phi}(\omega) = -\hat{q}(\omega)/\hat{c}(\omega) \qquad\qquad (9.33)$$

where $q(\omega)$, $c(\omega)$ are the quadrature and co-spectra, respectively [see equation (8.14)].

Thus, having estimated the cross-spectrum, equations (9.32) and (9.33) enable us to estimate the gain and phase of the linear system, whether or not there is added noise.

We can also use cross-spectral analysis to estimate the properties of the noise process. The discrete-time version of equation (9.28) is

$$Y_t = \sum_{k=0}^{\infty} h_k X_{t-k} + N_t \qquad\qquad (9.34)$$

For mathematical convenience, we again assume that

$$E(N_t) = E(X_t) = 0 \text{ so that } E(Y_t) = 0.$$

If we multiply both sides of (9.34) by Y_{t-m}, we find

$$Y_t Y_{t-m} = (\Sigma h_k X_{t-k} + N_t)(\Sigma h_k X_{t-m-k} + N_{t-m})$$

Taking expectations we find

$$\gamma_{yy}(m) = \sum_k \sum_j h_k h_j \gamma_{xx}(m - k + j) + \gamma_{nn}(m)$$

since (X_t) and (N_t) are assumed to be uncorrelated.

Taking Fourier transforms of both sides of this equation, we find

$$f_y(\omega) = H(\omega)\, \overline{H(\omega)}\, f_x(\omega) + f_n(\omega)$$

But $H(\omega)\, \overline{H(\omega)} = G^2(\omega)$

$$= C(\omega) f_y(\omega) / f_x(\omega)$$

so that

$$f_n(\omega) = f_y(\omega)[1 - C(\omega)] \qquad (9.35)$$

Thus an estimate of $f_n(\omega)$ is given by

$$\hat{f}_n(\omega) = \hat{f}_y(\omega)[1 - \hat{C}(\omega)].$$

Equation (9.35) also enables us to see that if there is no noise, so that there is a pure linear relation between X_t and Y_t, then $f_n(\omega) = 0$ and $C(\omega) = 1$ for all ω. On the other hand if $C(\omega) = 0$ for all ω, then $f_y(\omega) = f_n(\omega)$ and the output is not linearly related to the input. This confirms the point mentioned in Chapter 8 that the coherency, $C(\omega)$, measures the linear correlation between input and output at frequency ω.

The results of this section, not only show us how to identify a linear system by cross-spectral analysis, but also give further guidance on the interpretation of functions derived from the cross-spectrum, particularly the gain, phase and coherency. An example, involving a chemical process, is given by Goodman *et al.* (1961), while Gudmundsson (1971) describes an economic application of cross-spectral analysis.

In principle, estimates of the frequency response function of a linear system may be transformed to give estimates of the impulse response function (Jenkins and Watts, 1968, p. 444) but I do not recommend this. For instance example 2 of Section 8.2.1 appears to indicate that the sign of the phase may be used to indicate which series is 'leading the other. But

Hause (1971) has shown that for more complicated lagged models of the form

$$Y_t = \sum_{k=0}^{\infty} h_k X_{t-k}, \qquad (9.36)$$

which are called *distributed lag* models by economists, it becomes increasingly difficult to make inferences from phase estimates and Hause concludes that phase leads and lags will rarely provide economists with direct estimates of the *time* domain relationships that are of more interest to them.

9.4.2 The Box–Jenkins approach This section gives a brief introduction to the method proposed by G. E. P. Box and G. M. Jenkins for identifying a physically realizable linear system, in the time domain, in the presence of added noise. Further details can be found in Box and Jenkins (1968, 1970).

The input and output series are both differenced d times until both are stationary, and are also mean-corrected. The modified series will be denoted by $\{X_t\}$, $\{Y_t\}$, respectively. We want to find the impulse response function, $\{h_k\}$, of the system where

$$Y_t = \sum_{k=0}^{\infty} h_k X_{t-k} + N_t \qquad (9.37)$$

The 'obvious' way to estimate $\{h_k\}$ is to multiply through equation (9.37) by X_{t-m} and take expectations to give

$$\gamma_{xy}(m) = h_0 \gamma_{xx}(m) + h_1 \gamma_{xx}(m-1) + \ldots \qquad (9.38)$$

assuming that N_t is uncorrelated with the input. If we assume that the weights $\{h_k\}$ are effectively zero beyond $k = K$, then the first $(K + 1)$ equations of the type (9.38) for $m = 0, 1, \ldots,$ K, can be solved for the $(K + 1)$ unknowns $h_0, h_1, \ldots, h_K$, on substituting estimates of γ_{xy} and γ_{xx}. Unfortunately these equations do not, in general, provide good estimators for the $\{h_k\}$, and, in any case, assume knowledge of the

219

truncation point K. The basic trouble, as already noted in Section 8.1.2, is that autocorrelation within the input and output series will increase the variance of cross-correlation estimates.

Box and Jenkins (1970, Chapter 1) therefore propose two modifications to the above procedure. Firstly, they suggest 'prewhitening' the input before calculating the sample cross-covariance function. Secondly, they propose an alternative form of equation (9.36) which will in general require fewer parameters. They represent the linear system by the equation

$$Y_t - \delta_1 Y_{t-1} - \ldots - \delta_r Y_{t-r} = \omega_0 X_{t-b}$$
$$- \omega_1 X_{t-b-1} \ldots - \omega_s Y_{t-b-s}$$

$$(9.39)$$

This is rather like equation (9.4), but is given in the notation used by Box and Jenkins (1970) and involves an extra parameter, b, which is called the *delay* of the system. The delay can be any non-negative integer. Using the backward shift operator, B, (9.39) may be written as

$$\delta(B)Y_t = \omega(B)X_{t-b} \qquad (9.40)$$

where

$$\delta(B) = 1 - \delta_1 B - \ldots - \delta_r B^r$$

and

$$\omega(B) = \omega_0 - \omega_1 B - \ldots - \omega_s B^s.$$

Box and Jenkins (1970) describe equation (9.40) as a *transfer function* model, which is not a good description in that many authors use the term transfer function to describe the frequency response function [equation (9.10)].

The Box–Jenkins procedure begins by fitting an ARMA model to the (differenced) input. Suppose this model is of the form (see Section 3.4.5)

$$\phi(B)X_t = \theta(B)\alpha_t$$

where $\{\alpha_t\}$ denotes a purely random process, in the notation of Box and Jenkins. Thus we can transform the input to a white noise process by

$$\phi(B)\theta^{-1}(B)X_t = \alpha_t.$$

Suppose we apply the same transformation to the output, to give

$$\phi(B)\theta^{-1}(B)Y_t = \beta_t$$

and then calculate the cross-covariance function of the filtered input and output, namely (α_t) and (β_t). It turns out that this function gives a better estimate of the impulse response function, since, if we write

$$h(B) = h_0 + h_1 B + h_2 B^2 + \dots$$

so that

$$Y_t = h(B)X_t + N_t,$$

then

$$\begin{aligned} \beta_t &= \phi(B)\theta^{-1}(B)Y_t \\ &= \phi(B)\theta^{-1}(B)[h(B)X_t + N_t] \\ &= h(B)\,\alpha_t + \phi(B)\theta^{-1}(B)N_t \end{aligned}$$

and

$$\gamma_{\alpha\beta}(m) = h_m \text{ Var}(\alpha_t) \tag{9.41}$$

since (α_t) is a purely random process, and N_t is uncorrelated with (α_t). Equation (9.41) is of a much simpler form to equation (9.38). If we denote the sample cross-covariance function of α_t and β_t by $c_{\alpha\beta}$, and the observed variance of α_t by s_α^2, then an estimate of h_m is given by

$$\hat{h}_m = c_{\alpha\beta}(m)/s_\alpha^2. \tag{9.42}$$

These estimates should be more reliable than those given by the solution of equations of type (9.38).

Box and Jenkins (1970) give the theoretical impulse response functions for a variety of models of type (9.39) and go on to show how the shape of the estimated impulse response function given by (9.42) can be used to suggest appropriate values of the integers r, b, and s in equation (9.39). They then show how to obtain least squares estimates of δ_1, δ_2, . . . , ω_0, ω_1, . . . , given values of r, b, and s. These estimates can in turn be used to obtain refined estimates of (h_m) if desired.

Box and Jenkins go on to show how a model of type (9.39) can be used for forecasting and control. It is too early to assess the effectiveness of the method, but the few case studies which I have seen indicate that the method is potentially useful.

One general question which remains is whether it is 'best' to use cross-spectral analysis or the Box–Jenkins time-domain approach to identify a linear system. In order to set up a model for a gas furnace, Box and Jenkins (1970; p. 381) analyse the same data studied in the frequency domain by Jenkins and Watts (1968, p. 444) and it is suggested (Jenkins and Watts, 1968; p. 422 and p. 448) that, while cross-spectral analysis may be a useful diagnostic tool, spectral methods involve estimating a parameter at each frequency and more positive results may usually be obtained with a time-domain model, such as (9.39), which requires fewer parameters. But this gas furnace example has been criticized by Young (1974) on the grounds that the noise level is low so that a simple regression analysis may be adequate, and also because there is evidence of non-stationarity. Control engineers have also been concerned with comparing methods based on parametric models to frequency methods (e.g. Astrom and Eykhoff, 1970). Hamon and Hannan (1963) have suggested that spectral methods may be used to construct regression models. My own preference is to use frequency methods when interested in the frequency properties of a linear system, and time-domain methods when interested in a particular class of parametric models.

In conclusion, it is worth noting that a similar method to the Box–Jenkins approach has been independently developed by two control engineers, K. J. Astrom and T. Bohlin. This method also involves prewhitening and a model of type (9.39) (Astrom and Bohlin, 1966; Astrom, 1970), but does not discuss identification and estimation procedures in equal depth. One difference of the Astrom–Bohlin method is that non-stationary series may be converted to stationarity by methods other than differencing. The results of Clarke (1973) suggest that high-pass filtering or linear trend removal may sometimes be better than differencing.

9.4.3 Systems involving feedback

A system of the type illustrated in Fig. (9.7) is called an *open-loop* system, and the procedures described in the previous two sections are appropriate for data collected under these conditions. But data is often collected from systems where some form of *feedback control* is being applied, and then we have what is called a *closed-loop* system as illustrated in Fig. 9.8. For example, when trying to identify a full-scale industrial process, it could be dangerous, or an unsatisfactory product could be produced, unless some form of feedback control is applied to keep the output somewhere near target. Similar problems arise in an economic context. For example, attempts to find a linear relationship showing the effect of price changes on wage changes are bedevilled by the fact that wage changes will in turn affect prices.

The problem of identifying systems in the presence of feedback control is discussed by Granger and Hatanaka (1964, Chapter 7), Akaike (1967), Wilson (1971), Astrom and Eykhoff (1971, p. 130), Priestley (1969a, 1971), and Box and Macgregor (1974), and it is important to realise that open-loop procedures may not be applicable. The situation can be explained more clearly in the frequency domain. Let

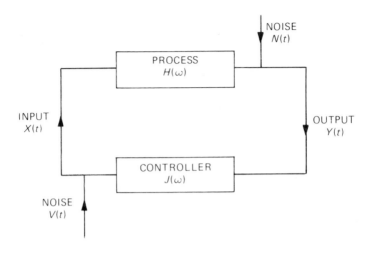

Figure 9.8   A closed-loop system.

$f_{xy}(\omega)$ denote the cross-spectrum of $X(t)$ and $Y(t)$ in
Fig. 9.8, and let $f_x(\omega)$, $f_n(\omega)$, $f_v(\omega)$ denote the spectra of
$X(t)$, $N(t)$, and $V(t)$ respectively. Then if $H(\omega)$ and $J(\omega)$
denote the frequency response functions of the system and
controller respectively, it can be shown (e.g. Akaike, 1967)
that

$$f_{xy}/f_x = (Hf_v + \bar{J}f_n)/(f_v + J\bar{J}f_n) \tag{9.43}$$

where all terms are functions of frequency, and $\bar{J}$ is the
complex conjugate of J. Only if $f_n \equiv 0$ or $J \equiv 0$, is the ratio
f_{xy}/f_x equal to H as is the case for an open-loop system
(equation 9.31). Thus the estimate of H provided by $\hat{f}_{xy}/\hat{f}_x$
will be poor unless f_n/f_v is small (Priestley, 1971). In
particular if $f_v \equiv 0$, $\hat{f}_{xy}/\hat{f}_x$ will provide an estimate of J^{-1} and
not of H.

Similar remarks apply to an analysis in the time domain.
The time-domain equivalent of (9.43) is given by Box and
MacGregor (1974).

The above problem is not specifically discussed by Box
and Jenkins (1970), although it is quite clear from the

224

remarks in their Section 11.6 that their methods are only intended for use in open-loop systems. However some confusion may be created by the fact that Box and Jenkins (1970; Section 12.2) do discuss ways of choosing optimal feedback control which is in fact quite a different problem. Having identified a system in open-loop, they show how to choose the feedback control action so as to satisfy some chosen criterion.

Unfortunately open-loop identification procedures have sometimes been used for a closed-loop system where they are not appropriate. Tee and Wu (1972) studied a paper machine while it was already operating under manual control and proposed a control procedure which has been shown to be worse than the existing form of control (Box and MacGregor, 1974). In a marketing situation, T.C.A. (1971) investigate the relationship between expenditure on offers, deals and media advertising on the sales of washing-up liquid and coffee. It is, however, often the case that expenditure on advertising is in turn affected by changes in sales levels, which may throw doubt on the conclusions contained in the above report.

What then can be done if feedback is present? Box and MacGregor (1974) suggest one possible approach in which one deliberately adds an independent programmed noise sequence on top of the noise, $V(t)$. Alternatively one may have some knowledge of the noise structure or of the controller frequency response function. Akaike (1968) claims that it is possible to identify a system provided only that instantaneous transmission of information does not occur in both system and controller, and an example of his, rather complicated, procedure is given by Otomo et al. (1972).

However a further difficulty is that it is not always clear if feedback is present or not, particularly in economics and marketing. Some indication may be given by the methods of Granger and Hatanaka (1964, Chapter 7), or by observing significantly large cross-correlation coefficients between (prewhitened) input and output at a zero or positive lag.

Alternatively it may be clear from physical considerations that feedback is or is not present. For example in studying the relationship between average (ambient) temperature and sales of a product, it is clear that sales cannot possibly affect temperature so that one has an open-loop system. But if feedback is present, then standard cross-correlation and cross-spectral analysis should not be used as they may give spurious results.

Exercises

1. Find the impulse response function, the step response function, the frequency response function, the gain and the phase shift for the following linear systems (or filters).

   (a) $\quad y_t = \frac{1}{2}x_{t-1} + x_t + \frac{1}{2}x_{t+1}$

   (b) $\quad y_t = \frac{1}{5}(x_{t-2} + x_{t-1} + x_t + x_{t+1} + x_{t+2})$

   (c) $\quad y_t = \nabla x_t$

   (d) $\quad y_t = \nabla^2 x_t$

   where in each case t is integer-valued. Plot the gain and phase shift for filters (a) and (c). Which of the filters are (i) low-pass (ii) high-pass?

   If the filters (a) and (b), given above, are joined in series, find the frequency response function of the combined filter.

2. Find the frequency response functions of the following linear systems in continuous time.

   (a) $\quad y(t) = gx(t - \tau)$

   where g, τ are positive constants

   (b) $\quad y(t) = \frac{g}{T} \int_0^\infty e^{-u/T} x(t - u) \, du$

226

3. If $\{X_t\}$ is a stationary discrete time-series with power spectral density function $f(\omega)$, show that the smoothed time-series

$$Y_t = \sum_{p=0}^{k} a_p X_{t-p},$$

where the a's are real constants, is a stationary process which has power spectral density function

$$\left[\sum_{q=0}^{k} \sum_{p=0}^{k} a_p a_q \cos(p - q)\omega \right] f(\omega).$$

In particular if $a_p = 1/(k + 1)$ for $p = 0, 1, \ldots, k$, show that the power spectrum of Y_t is

$$f(\omega)[1 - \cos(k + 1)\omega]/(k + 1)^2 (1 - \cos \omega).$$

[Hint: Use equation (9.21) and the trigonometric relation $\cos A \cos B = \frac{1}{2}[\cos(A + B) + \cos(A - B)]$.]

4. Consider the one-parameter second-order AR process

$$X_t = \alpha X_{t-2} + Z_t$$

where $\{Z_t\}$ denotes a purely random process, mean zero, variance σ_Z^2. Show that the process is second-order stationary if $|\alpha| < 1$, and find its autocovariance and autocorrelation functions.

Show that the power spectral density function of the process is given by

$$f(\omega) = \sigma_Z^2 / \pi(1 - 2\alpha \cos 2\omega + \alpha^2) \quad (0 < \omega < \pi)$$

using two different methods: a) By transforming the autocovariance function. b) By using the approach of Section 9.3.4.

Suppose now that $\{Z_t\}$ is any stationary process which has power spectrum $f_Z(\omega)$. What then is the power spectrum of $\{X_t\}$ as defined by the above equation?

CHAPTER 10

Some other Topics

This chapter provides a brief introduction and references for some topics not covered in the rest of the book.

CONTROL THEORY   Many books have been entirely devoted to this topic, which is growing rapidly. To many statisticians, 'control' implies statistical quality control using control charts. To control engineers, 'control' implies finding an automatic control procedure for a system whose structure may or may not be known. Control engineers have made substantial contributions to control theory in their search for optimal or near-optimal procedures. Contributions to control theory have also been made by mathematicians and operational researchers.

Control engineers were originally concerned only with deterministic control (and many of them still are). Some of the research, such as the solution of non-linear differential equations subject to boundary conditions, has an obvious relevance to control theory, but may equally be regarded as a branch of applied mathematics. Some of this work is also relevant to stochastic control where the system being controlled is subject to random disturbances. Some references are Douce (1963), Laning and Battin (1956), Solodovnikov (1965), Astrom (1970), Bell (1973), Fuller (1970) and Jacobs (1974).

In stochastic control, one wants to separate the signal from the noise (see Section 5.6) and there has been much work on filtering starting with the work of Wiener and Kolmogorov. A recent major development by control engineers is the Kalman filter which is a recursive method of estimating the state of a system in the presence of noise. The clearest introduction I have seen to this method is in the paper by Young (1974) who emphasizes the connection with some earlier statistical work by R. L. Plackett. Kalman filtering has been used in controlling the flight of a space rocket where the system dynamics are well defined but the disturbances are unknown.

The major statistical contribution to control theory has been the work of Box and Jenkins (1970). They show how to identify a suitable model for a system and hence find a 'good' control policy. A somewhat similar approach has been developed independently by Astrom and Bohlin (1965). MacGregor (1972) has compared the Box–Jenkins approach with the control engineer's state variable approach. MacGregor points out that there has been surprisingly little communication between statisticians and control engineers, even though some form of co-operation is clearly desirable. There are many differences in approach. For example Box and Jenkins stress identification, whereas control engineers often assume knowledge of all system and noise parameters, which may not be realistic. Box and Jenkins work mainly in discrete time using difference equations, whereas control engineers tend to work in continuous time using differential equations (although they do also use sampled-data control theory). Caines (1972) has compared the Box–Jenkins approach with Kalman filtering.

Some other statistical contributions to control theory are given in the symposium on control theory organized by the Royal Statistical Society in 1969. The first paper (Wishart, 1969) represents a mathematician's view of control theory and, apart from his Section 5.6, deals exclusively with deterministic control. The second paper (Whittle, 1969) deals with some aspects of stochastic control, particularly where the process is of a Markovian character. The third paper (Bather, 1969) describes how dynamic programming may be applied to the control of a Markov process. The mathematical level of the papers is very advanced, and they pay little attention to the problem of estimating the structure of a system, though Bather (1969) admits that 'the presence of unknown parameters is a serious threat to the methods he considers'. In the subsequent discussion, Priestley (1969b) pointed out that there is a considerable gap between the theoretical aspects of stochastic control discussed in the three papers and practical considerations. He suggested that the two aspects of control theory should be connected in a closed-loop system with each stimulating and reacting with the other, a view with which I entirely agree.

NON-STATIONARY PROCESSES We have stressed throughout the book that obvious sources of non-stationarity, such as trend in mean, should be removed from an observed time series before fitting a stationary model. Nevertheless the properties of the filtered series may still change with time, albeit slowly in some cases. It is therefore often a good idea to split the series into two or more reasonably long segments and compare the properties of each segment, particularly the form of the autocorrelation function and spectrum.

If the spectrum is thought to be changing through time, the 'evolutionary spectral approach' may be useful in which the spectral properties of the time series are examined for *overlapping* subsets. The term 'evolutionary spectrum' was introduced by Priestley (1965) to describe a particular form of time-dependent spectrum which changes *slowly* with time. Some recent papers which consider the problem of defining the spectrum for a non-stationary process are Loynes (1968), Priestley and Tong (1973) and Ackroyd (1973). Hammond (1973) has applied these ideas to the analysis of jet noise.

While on the subject of non-stationary processes, it is worth noting that the Kalman filtering technique for estimating the state of a system in the presence of noise (e.g.

Young, 1974) will work for certain types of non-stationary process as well as for stationary processes.

THE CEPSTRUM   This word, which is an anagram of spectrum, is the power spectrum of the logarithmic power spectrum of a given time series. If a time series contains an echo, with a constant delay time, then the analysis of a finite record will result in a periodic spectrum with peaks at the harmonics which are integer multiples of the fundamental frequency. A spectral analysis of such a spectrum may be a helpful step in the analysis. The term 'cepstrum' was invented by J. W. Tukey and introduced in a paper by Bogert *et al.* (1963). This paper also coined such terms as 'quefrency' and 'saphe' which have 'obvious' meanings. A recent reference is Noll (1973).

OBSERVATIONS AT UNEQUAL INTERVALS   This book has been mainly concerned with discrete time series measured at equal intervals of time. When observations are taken at unequal intervals the general definition of the periodogram (equation 7.17) still applies but the values of t are no longer integers. Autocorrelation coefficients can only be obtained directly by fitting a smoothed approximation to the time series. *Splines*, which are piecewise polynomials, can be used to provide such an approximation (see Wold, 1974). Further references are Hannan (1970, p. 48), Godfrey (1974) and Shapiro and Silverman (1960). Neave (1970) and Jones (1971) have considered the problem of analysing time series from which some observations are missing.

POLYSPECTRA   Brillinger (1965, 1975) discusses the higher-order spectra, or *polyspectra*, of multivariate stationary time series which are the Fourier transform of the appropriate cumulant. For a single time series, the first-order polyspectrum is the usual power spectrum, while the second-order polyspectrum is the *bispectrum*, discussed by

Godfrey (1965), which may be able to determine the form of a non-linear element. For a pair of time series, the first-order polyspectrum is the usual cross-spectrum. I have no practical experience with polyspectra and will not pursue the topic.

SPATIAL SERIES   Time series are observations ordered with respect to time. Time-series methods can also be used on series ordered with respect to some other variable, such as space. In geography, data often arise where the observations are ordered with respect to *two* spatial co-ordinates. Such data are called spatial series. The matrix of spatial autocorrelations can be regarded as a two-dimensional generalization of a time-series autocorrelation function. Methods of analysing spatial series are still being developed, but the methods are already proving useful (see for example Unwin and Hepple, 1974; Cliff and Ord, 1973; Besag, 1974).

'CROSSING' PROBLEMS   In a variety of practical problems, it is useful to have information about the 'level-crossing' properties of a time series. This concerns the times at which the series crosses the zero axis or some other arbitrary level. The problem is discussed by Cramer and Leadbetter (1967), and a recent comprehensive survey and bibliography is given by Blake and Lindsey (1973).

The Fourier, Laplace and Z-transforms

This appendix provides a brief introduction to the Fourier transform which is a valuable mathematical tool in time-series analysis. Further details may be found in many books such as Hsu (1967) and Sneddon (1972). The related Laplace and Z-transforms are also introduced.

Given a (possibly complex-valued) function $h(t)$ of a real variable t, the Fourier transform of $h(t)$ is usually defined as

$$H(\omega) = \int_{-\infty}^{\infty} h(t)e^{-i\omega t}\,dt \tag{A1.1}$$

provided the integral exists for every real ω. Note that $H(\omega)$ is in general complex. A sufficient condition for $H(\omega)$ to exist is

$$\int_{-\infty}^{\infty} |h(t)|\,dt < \infty.$$

If (A1.1) is regarded as an integral equation for $h(t)$ given $H(\omega)$, then a simple inversion formula exists of the form

$$h(t) = \frac{1}{2\pi} \int_{-\infty}^{\infty} H(\omega)e^{i\omega t}\,d\omega \tag{A1.2}$$

and $h(t)$ is called the inverse Fourier transform of $H(\omega)$, or sometimes just the Fourier transform of $H(\omega)$. The two functions $h(t)$ and $H(\omega)$ are commonly called a Fourier transform pair.

The reader is warned that many authors use a slightly different definition of a Fourier transform to A1.1. For example some authors put a constant $1/\sqrt{2\pi}$ outside the integral in (A1.1) and then the inversion formula for $h(t)$ is symmetric. In time-series analysis many authors (e.g. Cox and Miller, 1968, p. 315) put a constant $1/2\pi$ outside the integral in A1.1. The inversion formula then has a unity constant outside the integral.

Some authors in time-series analysis (e.g. Jenkins and Watts, 1968, Blackman and Tukey, 1959) define Fourier transforms in terms of the variable $f = \omega/(2\pi)$ rather than ω. We then find that the Fourier transform pair is

$$G(f) = \int_{-\infty}^{\infty} h(t)e^{-2\pi ift}dt \qquad (A1.3)$$

$$h(t) = \int_{-\infty}^{\infty} G(f)e^{2\pi ift}df \qquad (A1.4)$$

Note that the constant outside each integral is now unity.

In time-series analysis, we will often use the discrete form of the Fourier transform when $h(t)$ is only defined for integer values of t. Then

$$H(\omega) = \sum_{t=-\infty}^{\infty} h(t)e^{-i\omega t} \qquad (-\pi \leqslant \omega \leqslant \pi) \qquad (A1.5)$$

is the discrete Fourier transform of $h(t)$. Note that $H(\omega)$ is only defined in the interval $[-\pi, \pi]$. The inverse transform is

$$h(t) = \frac{1}{2\pi}\int_{-\pi}^{\pi} H(\omega)e^{i\omega t}d\omega \qquad (A1.6)$$

Fourier transforms have many useful properties, some of which are used during the later chapters of this book.

THE FOURIER, LAPLACE AND Z-TRANSFORMS

However we will not attempt to review them here. The reader is referred for example to Hsu (1967).

One special type of Fourier transform arises when $h(t)$ is a real-valued even function such that $h(t) = h(-t)$. The autocorrelation function of a stationary time series has these properties. Then using (A1.1) with a constant $1/\pi$ outside the integral, we find

$$H(\omega) = \frac{1}{\pi} \int_{-\infty}^{\infty} h(t)e^{-i\omega t}\mathrm{d}t$$

$$= \frac{2}{\pi} \int_{0}^{\infty} h(t) \cos \omega t \, \mathrm{d}t \qquad \text{(A1.7)}$$

and it is clear that $H(\omega)$ is a real–valued even function. The inversion formula is then

$$h(t) = \frac{1}{2} \int_{-\infty}^{\infty} H(\omega)e^{i\omega t}\mathrm{d}\omega \qquad \text{(A1.8)}$$

$$= \int_{0}^{\infty} H(\omega)\cos \omega t \, \mathrm{d}\omega$$

Equations (A1.7) and (A1.8) are similar to a Fourier transform pair and are useful when we only wish to define $H(\omega)$ for $\omega > 0$. This pair of equations appear as equations (2.73) and (2.74) in Yaglom (1962). When $h(t)$ is only defined for integer values of t, equations (A1.7) and (A1.8) become

$$H(\omega) = \frac{1}{\pi}\left\{h(0) + 2 \sum_{t=1}^{\infty} h(t) \cos \omega t\right\} \qquad \text{(A1.9)}$$

$$h(t) = \int_{0}^{\pi} H(\omega) \cos \omega t \, \mathrm{d}\omega \qquad \text{(A1.10)}$$

and $H(\omega)$ is now only defined on $[0, \pi]$.

The Laplace transform of a function $h(t)$ which is defined

for $t > 0$ is given by

$$H(s) = \int_0^\infty h(t)e^{-st}\, dt \qquad\qquad (A1.11)$$

where s is a complex variable. The integral converges when the real part of s exceeds some number called the abscissa of convergence.

The relationship between the Fourier and Laplace transforms is of some interest, particularly as control engineers often prefer to use the Laplace transform when investigating the properties of a linear system (e.g. Solodovnikov, 1965, Douce, 1963) as this will cope with physically realizable systems which are stable *or* unstable. If the function $h(t)$ is such that

$$h(t) = 0 \qquad (t < 0) \qquad\qquad (A1.12)$$

and the real part of s is zero, then the Laplace and Fourier transforms of $h(t)$ are the same. The impulse response function of a physically realizable linear system satisfies (A1.12) and so for such functions the Fourier transform is a special case of the Laplace transform. More details about the Laplace transform may be found in many books, (e.g. Sneddon, 1972).

The Z-transform of a function $h(t)$ defined on the non-negative integers is given by

$$H(z) = \sum_{t=0}^\infty h(t)z^{-t} \qquad\qquad (A1.13)$$

In discrete time, with a function satisfying (A1.12), some authors prefer to use the Z-transform rather than the discrete form of the Fourier transform (i.e. A1.5) or the discrete form of the Laplace transform, namely

$$H(s) = \sum_{t=0}^\infty h(t)e^{-st} \qquad\qquad (A1.14)$$

Comparing (A1.13) with (A1.14) we see that $z = e^s$. The reader will observe that when $\{h(t)\}$ is a probability function such that $h(t)$ is the probability of observing the value t, for $t = 0, 1, \ldots$, then (A1.13) is related to the probability generating function of the distribution, while (A1.5) and (A1.14) are related to the moment generating function.

Exercises

1. If $h(t)$ is real, show that the real and imaginary parts of its Fourier transform, as defined by equation (A1.1), are even and odd functions respectively.

2. If $h(t) = e^{-a|t|}$ for all real t, where a is a positive real constant, show that its Fourier transform, as defined by equation (A1.1), is given by

$$H(\omega) = 2a/(a^2 + \omega^2) \qquad (-\infty < \omega < \infty)$$

3. Show that the Laplace transform of $h(t) = e^{-at}$ $(t > 0)$, where a is a real constant, is given by

$$H(s) = 1/(s + a) \qquad \text{Re}(s) > a.$$

The Dirac Delta Function

Suppose that $\phi(t)$ is any function which is continuous at $t = 0$. Then the *Dirac delta function,* $\delta(t)$, is such that

$$\int_{-\infty}^{\infty} \delta(t)\phi(t)\mathrm{d}t = \phi(0) \tag{A2.1}$$

Because it is defined in terms of its integral properties alone, it is sometimes called the 'spotting' function since it picks out one particular value of $\phi(t)$. It is also sometimes called simply the *delta function.*

It is important to realize that $\delta(t)$ is *not* a function. Rather it is a generalized function, or distribution, which maps a function into the real line.

Some authors define the delta function by

$$\delta(t) = \begin{cases} 0 & t \neq 0 \\ \infty & t = 0 \end{cases} \tag{A2.2}$$

such that

$$\int_{-\infty}^{\infty} \delta(t)\mathrm{d}t = 1.$$

But while this is often intuitively helpful, it is mathematically meaningless.

The Dirac delta function can also be regarded (e.g. Schwarz and Friedland, 1965) as the limit, as $\epsilon \to 0$, of a pulse of width ϵ and height $1/\epsilon$ (i.e. unit area) defined by

$$u(t) = \begin{cases} 1/\epsilon & 0 < t < \epsilon \\ 0 & \text{otherwise} \end{cases}$$

THE DIRAC DELTA FUNCTION

This definition is also not mathematically rigorous, but heuristically useful. In particular, control engineers can approximate such an impulse by an impulse with unit area whose duration is short compared with the least significant time constant of the response to the linear system being studied.

Even though $\delta(t)$ is a generalized function, it can often be handled as if it were an ordinary function except that we will be interested in the values of integrals involving $\delta(t)$ and never in the value of $\delta(t)$ by itself.

The derivative $\delta'(t)$ of $\delta(t)$ can also be defined by

$$\int_{-\infty}^{\infty} \delta'(t)\phi(t)\mathrm{d}t = -\phi'(0) \tag{A2.3}$$

where $\phi'(0)$ is the derivative of $\phi(t)$ evaluated at $t = 0$. The justification for A2.3 depends on integrating by parts as if $\delta'(t)$ and $\delta(t)$ were ordinary functions and using (A2.2) to give

$$\int_{-\infty}^{\infty} \delta'(t)\phi(t)\mathrm{d}t = -\int_{-\infty}^{\infty} \delta(t)\phi'(t)\mathrm{d}t$$

and then using (A2.1). Higher derivatives of $\delta(t)$ may be defined in a similar way.

The delta function has many useful properties (see, for example, Hsu, 1967).

Exercises

1. The function $\phi(t)$ is continuous at $t = t_0$. If $a < b$, show that

$$\int_a^b \delta(t - t_0)\phi(t)\mathrm{d}t = \begin{cases} \phi(t_0) & a < t_0 < b \\ 0 & t_0 < a, t_0 > b. \end{cases}$$

2. The function $\phi(t)$ is continuous at $t = 0$. Show that $\phi(t)\delta(t) = \phi(0)\delta(t)$.

239

APPENDIX III

Covariance

Any reader who is unfamiliar with the laws of probability, independence, probability distributions, mean, variance and expectation, and elementary statistical inference should consult an elementary book on statistics, such as Freund, J. E. (1962), Mathematical Statistics, Prentice-Hall.

The idea of covariance is particularly important in the study of time series, and will now be briefly revised.

Suppose two random variables, X and Y, have means μ_x, μ_y, respectively. Then the covariance of X and Y is defined to be

$$\text{Cov}(X,Y) = E\{(X - \mu_x)(Y - \mu_y)\}.$$

If X and Y are independent, then

$$E[(X - \mu_x)(Y - \mu_y)] = E(X - \mu_x)E(Y - \mu_y)$$
$$= 0$$

so that the covariance is zero. If X and Y are *not* independent, then the covariance may be positive or negative depending on whether 'high' values of X tend to go with 'high' or 'low' values of Y.

Covariance is a useful quantity for many mathematical purposes, but it is difficult to interpret as it depends on the units in which X and Y are measured. Thus it is often useful to standardize the covariance between two random variables by dividing by the product of their respective standard deviations to give a quantity called the *correlation*

240

coefficient. It can easily be shown that the correlation coefficient must lie between ± 1.

If X and Y are random variables from the same stochastic process at different times, then the covariance coefficient is called an *auto*covariance coefficient, and the correlation coefficient is called an *auto*correlation coefficient. If the process is stationary, the standard deviations of X and Y will be the same and their product will be the variance of X (or of Y).

References

ACKROYD, M. H. (1973) *Time-dependent spectra: the unified approach. In* Griffiths and Stocklin (1973).

AKAIKE, H. (1967) *Some problems in the application of the cross-spectral method. In* Harris (1967).

AKAIKE, H. (1968) On the use of a linear model for the identification of feedback systems, *Annals of the Inst. Statist. Math., 20,* 425–439.

AKAIKE, H. (1973a) Maximum likelihood identification of Gaussian auto-regressive moving average models, *Biometrika, 60,* 255–266.

AKAIKE, H. (1973b) Contribution to the discussion of Chatfield and Prothero (1973). *J. R. Statist. Soc. A, 136,* 330.

ANDERSON, T. W. (1971) *The Statistical Analysis of Time Series.* New York: Wiley.

ARMSTRONG, J. S. and GROHMAN, M. C. (1972) A comparative study of methods for long-range forecasting, *Man. Sci., 19,* 211–221.

ASTROM, K. J. (1970) *Introduction to Stochastic Control Theory.* New York: Academic Press.

ASTROM, K. J. and BOHLIN, T. (1966) Numerical identification of linear dynamic systems from normal operating records, *In Theory of Self-Adaptive Control Systems,* ed. P. M. Hammond, pp. 94–111, New York: Plenum Press.

ASTROM, K. J. and EYKHOFF, P. (1971) System identification – a survey, *Automatica, 7,* 123–162.

BARKSDALE, H. C. and GUFFEY, H. J. Jr., (1972) An illustration of cross-spectral analysis in marketing, *J. of Marketing Research, 9,* 271–278.

BARNARD, G. A., JENKINS, G. M. and WINSTEN, C. B. (1962) Likelihood, inference and time-series, *J. R. Statist. Soc., A, 125,* 321–372.

BARTLETT, M. S. (1966) *Stochastic Processes,* 2nd edn. Cambridge Univ. Press.

REFERENCES

- BARTLETT, M. S. (1967) Some remarks on the analysis of time-series, *Biometrika*, **54**, 25–38.
- BATES, J. M. and GRANGER, C. W. J. (1969) The combination of forecasts, *Op. Res. Quart.*, **20**, 451–468.
- BATHER, J. A. (1969) Diffusion models in stochastic control theory, *J. R. Statist. Soc.*, A, **132**, 335–345.
- BELL, D. J. (1973) Recent Mathematical Developments in Control., *Proc. of a conference at Bath University.* London: Academic Press.
- BENDAT, J. S. and PIERSOL, A. G. (1971) *Random Data: Analysis and Measurement Procedures.* 2nd edn. New York: Wiley.
- BESAG, J. (1974) Spatial interaction and the statistical analysis of lattice systems, *J. Roy. Statist. Soc.* B, **36**, 192–236.
- BHATTACHARYYA, M. N. (1974) Forecasting the demand for telephones in Australia, *Appl. Statist.*, **23**, 1–10.
- BINGHAM, C., GODFREY, M. D. and TUKEY, J. W. (1967) Modern techniques of power spectrum estimation, *IEEE Trans. on Audio and Electroacoustics,* AU–15, No. 2, 56–66.
- BLACKMAN, R. B. and TUKEY, J. W. (1959) *The Measurement of Power Spectra.* New York: Dover.
- BLAKE, I. F. and LINDSEY, W. C. (1973) Level-crossing problems for random processes, *IEEE Trans.*, Vol. IT-19, No. 3, 295–315.
- BLOOMFIELD, P. (1973) An exponential model for the spectrum of a scalar time-series, *Biometrika*, **60**, 217–226.
- BOGERT, B. P., HEALY, M. J. R. and TUKEY, J. W. (1963) The quefrency alanysis of time-series for echoes: cepstrum, pseudo-autocovariance, cross-cepstrum and saphe cracking, *In* Rosenblatt (1963).
- BOX, G. E. P. and COX, D. R. (1964) An analysis of transformations, *J. R. Statist. Soc.*, B, **26**, 211–252.
- BOX, G. E. P. and JENKINS, G. M. (1962) Some statistical aspects of adaptive optimisation and control, *J. R. Statist. Soc.*, B, **24**, 297–343.
- BOX, G. E. P. and JENKINS, G. M. (1968) Some recent advances in forecasting and control. Part I, *Appl. Statist.*, **17**, 91–109.
- BOX, G. E. P. and JENKINS, G. M. (1970) *Time-Series Analysis, Forecasting and Control.* San Francisco: Holden-Day.
- BOX, G. E. P. and MACGREGOR, J. F. (1974) The analysis of closed-loop dynamic-stochastic systems, *Technometrics,* **16**, 391–398.
- BOX, G. E. P. and NEWBOLD, P. (1971) Some comments on a paper of Coen, Gomme and Kendall, *J. R. Statist. Soc.* A., **134**, 229–240.
- BOX, G. E. P. and PIERCE, D. A. (1970) Distribution of residual

auto-correlations in autoregressive-integrated moving average time-series models, *J. Amer. Statist. Ass.*, **65**, 1509–1526.

BRASS, W. (1974) Perspective in population prediction: Illustrated by the statistics of England and Wales, *J. R. Statist. Soc.*, A, **137**, 532–583.

BRILLINGER, D. R. (1965) An introduction to polyspectra, *Ann. Math. Statist.*, **36**, 1351–1374.

BRILLINGER, D. R. (1973) A power spectral estimate which is insensitive to transients, *Technometrics*, **15**, 559–562.

BRILLINGER, D. R. (1975) *Time Series: Data Analysis and Theory*. New York: Holt, Rinehart and Winston.

BROWN, R. G. (1963) *Smoothing, Forecasting and Prediction*. Englewood Cliffs, N.J: Prentice-Hall.

BUTTON, K. J. (1974) A critical review of techniques used for forecasting car ownership in discrete areas, *The Statistician*, **23**, 117–128.

CAINES, P. E. (1972) Relationship between Box–Jenkins–Astrom control and Kalman linear regulator, *Proc. IEE*, **119**, 615–620.

CHAMBERS, J. C., MULLICK, S. K. and SMITH, D. D. (1971) How to choose the right forecasting technique, *Harvard Business Review*, **49**, No. 4, 45–74.

CHATFIELD, C. (1974) Some comments on spectral analysis in marketing', *J. of Marketing Research*, **11**, 97–101.

CHATFIELD, C. and NEWBOLD, P. (1974) Adaptive filtering, *Op. Res. Quart.*, **25**, 494–495.

CHATFIELD, C. and PEPPER, M. P. G. (1971) Time-series analysis: an example from geophysical data, *Appl. Statist.*, **20**, 217–238.

CHATFIELD, C. and PROTHERO, D. L. (1973a) Box–Jenkins seasonal forecasting: problems in a case-study, *J. R. Statist. Soc.* A, **136**, 295–336.

CHATFIELD, C. and PROTHERO, D. L. (1973b) A reply to some comments by Box and Jenkins, *J. R. Statist. Soc.* A, **136**, 345–352.

CHRIST, C. F. (1966) *Econometric Models and Methods*. New York: Wiley.

CLARKE, D. W. (1973) Experimental comparison of identification methods, Paper presented to UKAC 5th control convention, organized by Inst. of Measurement and Control.

CLIFF, A. D. and ORD, J. K. (1973) *Spatial Autocorrelation*. London: Pion.

COEN, P. J., GOMME, E. J. and KENDALL, M. G. (1969) Lagged relationships in economic forecasting, *J. R. Statist. Soc.* A, **132**, 133–163.

COOLEY, J. W., LEWIS, P. A. W. and WELCH, P. D. (1969) Historical

REFERENCES

notes on the Fast Fourier Transform, *IEEE Trans. On Aud. and Elect.,* AU-15, No. 2, 76–79.

COOLEY, J. W., LEWIS, P. A. W. and WELCH, P. D. (1969) The finite Fourier transform, *IEEE Trans. on Audio and Electroacoustics,* AU-17, No. 2, 77–85.

COX, D. R. (1961) Prediction by exponentially weighted moving averages and related methods, *J. Roy, Statist. Soc. B.* **23**, 414–422.

COX, D. R. and LEWIS, P. A. W. (1966) *The Statistical Analysis of Series of Events.* London: Methuen.

COX, D. R. and MILLER, H. D. (1968) *The Theory of Stochastic Processes.* New York: Wiley.

COUTIE, G. A. *et al.* (1964) *Short-term Forecasting, I.C.I. monograph* No. 2. Edinburgh: Oliver and Boyd.

CRADDOCK, J. M. (1965) The analysis of meteorological time series for use in forecasting, *The Statistician,* **15**, 169–190.

CRAMER, H. and LEADBETTER, M. R. (1967) *Stationary and Related Stochastic Processes.* New York: Wiley.

DAVIS, E. J. (1974) Forecasting in marketing, Proc. 1974 ESOMAR conference on forecasting in marketing.

DHRYMES, P. J. (1971) *Distributed Lags.* San Francisco: Holden-Day.

DOUCE, J. L. (1963) *The Mathematics of Servomechanisms.* London: E.U.P.

DURBIN, J. (1963) Trend elimination for the purpose of estimating seasonal and periodic components of time-series', *In* Rosenblatt (1963).

DURBIN, J. (1969) Tests for serial correlation in regression analysis based on the periodogram of least squares residuals, *Biometrika,* **56**, 1–16.

DURBIN, J. (1970) Testing for serial correlation in least-squares regression when some of the regressors are lagged dependent variables, *Econometrica,* **38**, 410–421.

DURBIN, J. (1973) Tests of model specification based on residuals. Symposium on Statistical Design and Linear Models, Fort Collins.

DURBIN, J. and WATSON, G. S. (1950; 1951; 1971) Testing for serial correlation in least-squares regression, I, II and III, *Biometrika,* **37**, 409–428, **38**, 159–178, **58**, 1–19.

FISHMAN, G. S. (1969) *Spectral Methods in Econometrics.* Cambridge, Mass.: Harvard Univ. Press.

FULLER, A. T. (1970) *Non-linear Stochastic Control Systems.* Taylor and Francis.

GEARING, H. W. G. (1971) Forecasting demand, *The Statistician,* **20**, 23–28.

GODFREY, M. D. (1965) An exploratory study of the bi-spectrum of economic time-series, *Appl. Statist.*, **14**, 48–69.

GODFREY, M. D. (1974) Computational methods for time series, *Bull. I.M.A.*, **10**, 224–227.

GOODMAN, N. R., KATZ, S., KRAMER, B. H. and KUO, M. T. (1961) Frequency response from stationary noise, *Technometrics*, **3**, 245–268.

GRANGER, C. W. J. (1966) The typical shape of an econometric variable, *Econometrica*, **34**, 150–161.

GRANGER, C. W. J. and HATANAKA, M. (1964) *Spectral Analysis of Economic Time Series*. Princeton: Princeton Univ. Press.

GRANGER, C. W. J. and HUGHES, A. O. (1968) Spectral analysis of short series – a simulation study, *J. R. Statist. Soc.* A, **131**, 83–99.

GRANGER, C. W. J. and HUGHES, A. O. (1971) A new look at some old data: the Beveridge wheat price series, *J. R. Statist. Soc.* A, **134**, 413–428.

GRANGER, C. W. J. and NEWBOLD, P. (1974) Spurious regressions in econometrics, *J. of Econometrics*, **2**, 111–120.

GREGG, J. V., HOSSELL, C. V. and RICHARDSON, J. T. (1964) *Mathematical Trend Curves: An Aid to Forecasting. I.C.I. Monograph* no. 1. Edinburgh: Oliver and Boyd.

GRENANDER, U. and ROSENBLATT, M. (1957) *Statistical Analysis of Stationary Time Series*. New York: Wiley.

GRIFFITHS, J. W. R. and STOCKLIN, P. L. eds. (1973) *Signal Processing*. London: Academic Press.

GUDMUNDSSON, G. (1971) Time-series analysis of imports, exports and other economic variables, *J. R. Statist. Soc.* A, **134**, 383–412.

HAMMOND, J. K. (1973) Evolutionary spectra in random vibrations, *J. R. Statist. Soc.* B, **35**, 167–188.

HAMON, B. V. and HANNAN, E. J. (1963) Estimating relations between time-series, *J. of Geophysical Research,* **68**, No. 21, 6033–6041.

HAMON, B. V. and HANNAN, E. J. (1974) Spectral estimation of time delay for dispersive and non-dispersive systems, *Appl. Statist.*, **23**, 134–142.

HANNAN, E. J. (1960) *Time Series Analysis*. London: Methuen.

HANNAN, E. J. (1970) *Multiple Time Series*. New York: Wiley.

HARRIS, B. (ed.) (1967) *Spectral Analysis of Time Series*. New York: Wiley.

HARRISON, P. J. (1965) Short-term sales forecasting, *Appl. Statist.*, **14**, 102–139.

REFERENCES

HARRISON, P. J. and PEARCE, S. F. (1972) The use of trend curves as an aid to market forecasting, *Industrial Marketing Management,* **2**, 149–170.

HARRISON, P. J. and STEVENS, C. F. (1971) A Bayesian approach to short-term forecasting, *Op. Res. Quart.,* **22**, 341–362.

HAUBRICH, R. A. and MACKENZIE, G. S. (1965) Earth noise, 5 to 500 millicycles per second, *J. of Geophysical Research,* **70**, No. 6, 1429–1440.

HAUSE, J. C. (1971) Spectral analysis and the detection of lead-lag relations, *Amer. Econ. Rev.,* **61**, 213–217.

HSU, H. P. (1967) *Fourier Analysis.* New York: Simon and Schuster.

JACOBS, O. L. R. (1974) *Introduction to Control Theory.* Oxford: Clarendon Press.

JENKINS, G. M. (1965) A survey of spectral analysis, *Appl. Statist.,* **14**, 2–32.

JENKINS, G. M. and WATTS, D. G. (1968) *Spectral Analysis and its Applications.* San Francisco: Holden-Day.

JONES, R. H. (1964) Spectral analysis and linear prediction of meteorological time series, *J. Applied Meteorology,* **3**, 45–52.

JONES, R. H. (1965) A re-appraisal of the periodogram in spectral analysis, *Technometrics,* **7**, 531–542.

JONES, R. H., (1971) Spectrum estimation with missing observations, *Annals of the Inst. Stat. Math.,* **23**, 387–398.

KEAY, F. (1972) *Marketing and Sales Forecasting.* London: Pergamon.

KENDALL, M. G. (1973) *Time-Series.* London: Griffin.

KENDALL, M. G. and STUART, A. (1966) *The Advanced Theory of Statistics,* Vol. 3. London: Griffin.

LANING, J. H. and BATTIN, R. H. (1956) *Random Processes in Automatic Control.* New York: McGraw-Hill.

LEE, Y. W. (1960) *Statistical Theory of Communication.* New York: Wiley.

LEUTHOLD, R. M., MACCORMICK, A. J. A., SCHMITZ, A. and WATTS, D. G. (1970) Forecasting daily hog prices and quantities: a study of alternative forecasting techniques. *J. Amer. Stat. Ass.,* **65**, 90–107.

LOYNES, R. M. (1968) On the concept of the spectrum for non-stationary processes, *J. R. Statist. Soc.* B, **30**, 1–30.

MACGREGOR, J. F. (1972) State variable approach to time series representation and forecasting, Technical report No. 306. Dept. of Stats., Univ. of Wisconsin.

MACKAY, D. B. (1973) A spectral analysis of the frequency of supermarket visits, *J. of Marketing Research,* **10**, 84–90.

MOOERS, C. N. K. and SMITH, R. L. (1968) Continental shelf waves off Oregon, *J. of Geophysical Research*, **73**, No. 2, 549–557.

MUNK, W. H., SNODGRASS, F. E. and TUCKER, M. J. (1959) Spectra of low-frequency ocean waves, *Bull. Scripps Inst. Oceanog.*, **7**, 283–362.

MYERS, C. L. (1971) Forecasting electricity sales, *The Statistician*, **20**, 15–22.

NAYLOR, T. H., SEAKS, T. G. and WICHERN, D. W. (1972) Box–Jenkins methods: an alternative to econometric models, *Int. Statist. Rev.*, **40**, 123–137.

NEAVE, H. R. (1970) Spectral analysis of a stationary time-series using initially scarce data, *Biometrika*, **57**, 111–122.

NEAVE, H. R. (1971) The exact error in spectrum estimators, *Ann. Math. Statist.*, **42**, 961–975.

NEAVE, H. R. (1972a) Observations on 'Spectral analysis of short series – a simulation study' by Granger and Hughes, *J. R. Statist. Soc.* A, **135**, 393–405.

NEAVE, H. R. (1972b) A comparison of lag window generators, *J. Amer. Statist. Ass.*, **67**, 152–158.

NELSON, C. R. (1973) *Applied Time Series Analysis for Managerial Forecasting.* San Francisco: Holden-Day.

NERLOVE, M. (1964) Spectral analysis of seasonal adjustment procedures, *Econometrica*, **32**, 241–286.

NERLOVE, M. and WALLIS, K. F. (1966) Use of the Durbin–Watson statistic in inappropriate situations, *Econometrica*, **34**, 235–238.

NEWBOLD, P. and GRANGER, C. W. J. (1974) Experience with forecasting univariate time-series and the combination of forecasts, *J. R. Statist. Soc.* A, **137**, 131–165.

NOLL, A. M. (1973) The cepstrum and some close relatives, *In* Griffiths and Stocklin (1973).

OTNES, R. K. and ENOCHSON, L. (1972) *Digital Time Series Analysis.* New York: Wiley.

OTOMO, T., NAKAGAWA, T. and AKAIKE, H. (1972) Statistical approach to computer control of cement rotary kilns, *Automatica*, **8**, 35–48.

PAPOULIS, A. (1965) *Probability, Random Variables, and Stochastic Processes.* New York: McGraw-Hill.

PAGANO, M. (1972) An algorithm for fitting autoregressive schemes, *Appl. Statist.*, **21**, 274–281.

PARZEN, E. (1962) *Stochastic Processes.* San Francisco: Holden-Day.

PARZEN, E. (1964) An approach to empirical time series analysis, *Radio Science J. of Research*, **68D**, 937–951, reprinted in Parzen (1967).

REFERENCES

PARZEN, E. (1967) *Time Series Analysis Papers*. San Francisco: Holden-Day.

PIERCE, D. A. (1970) A duality between autoregressive and moving average processes concerning their least squares parameter estimates, *Ann. Math. Statist.*, **41**, 722–726.

POCOCK, S. J. (1974) Harmonic analysis applied to seasonal variations in sickness absence, *Appl. Statist.*, **23**, 103–120.

PRIESTLEY, M. B. (1965) Evolutionary spectra and non-stationary processes. *J. Roy. Statist, Soc.* B, **27**, 204–237.

PRIESTLEY, M. B. (1969a) Estimation of transfer functions in closed loop stochastic systems, *Automatica*, **5**, 623–632.

PRIESTLEY, M. B. (1969b) Contribution to the discussion of Wishart (1969), Whittle (1969) and Bather (1969). *J. Roy. Statist. Soc.* A, **132**, 346–9.

PRIESTLEY, M. B. (1971) Fitting relationships between time-series, *Bull. of the Int. Statist. Inst.*, **44**, Book 1, 295–324.

PRIESTLEY, M. B. and TONG, H. (1973) On the analysis of bivariate non-stationary processes, *J. R. Statist. Soc.* B, **35**, 153–166.

QUENOUILLE, M. H. (1957) *The Analysis of Multiple Time Series*. London: Griffin.

REID, D. J. (1972) A comparison of forecasting techniques on economic time-series, Proc. of the 1971 conference on Forecasting in Action organised by the O.R. Soc. and the Soc. for Long Range Planning.

ROBINSON, E. A. (1967) *Multichannel Time Series with Digital Computer Programs*. San Francisco: Holden-Day.

RODEN, G. I. (1966) Low frequency sea level oscillations along the Pacific coast of N. America, *J. of Geophysical Research*, **71**, No. 20, 4755–4776.

ROSENBLATT, M. (ed) (1963) *Proceedings of Symposium on Time Series Analysis held at Brown University*, June 11–14, 1962. New York: Wiley.

SCHWARZ, R. J. and FRIEDLAND, B. (1965) *Linear Systems*. New York: McGraw-Hill.

SHAPIRO, H. S. and SILVERMAN, R. A. (1960) Alias-free sampling of random noise, *J. SIAM*, **8**, 225–248.

SHISKIN, J., YOUNG, A. M. and MUSGRAVE, J. C. (1967) The X-11 variant of the census method II seasonal adjustment program. Technical paper No. 15 (revised). Washington D.C.: U.S. Bureau of the Census.

SINGLETON, R. C. (1969) An algorithm for computing the mixed radix fast Fourier transform, *IEEE Trans. on Audio and Elect.*, Au-17, 93–103.

SNEDDON, I. H. (1972) *The Use of Integral Transforms.* New York: McGraw Hill.

SNODGRASS, F. E., GROVES, G. W., HASSELMANN, K. F., MILLER, G. R., MUNK, W. H. and POWERS, W. H. (1966) Propagation of ocean swell across the Pacific, *Phil. Trans. Roy. Soc., London,* A, **259**, 431–497.

SOLODOVNIKOV, V. V. (1965) *Statistical Dynamics of Linear Automatic Control Systems.* London: Van Nostrand.

TEE, L. H. and WU, S. M. (1972) An application of stochastic and dynamic models for the control of a papermaking process, *Technometrics,* **14**, 481–496.

TELEVISION CONSUMER AUDIT (1971) *Consumer response to promotional activity.* London: Television Consumer Audit.

TETLEY, H. (1946) *Actuarial Statistics,* Vol. 1. Cambridge: Cambridge Univ. Press.

THOMPSON, H. E. and TIAO, G. C. (1971) An analysis of telephone data: a case study of forecasting seasonal time-series, *Bell J. of Economics and Man. Sci.,* **2**, 515–541.

TICK, L. J. (1967) Estimation of coherency, In Harris (1967).

TOMASEK, O. (1972) Statistical forecasting of telephone time-series, *ITU Telecomm. J.,* December, 1–7.

TUKEY, J. W. (1967) An introduction to the calculations of numerical spectrum analysis. In Harris (1967).

UNWIN, D. J. and HEPPLE, L. W. (1974) The statistical analysis of spatial series, *The Statistician,* **23**, 211–227.

WAGLE, B., RAPPOPORT, J. Q. G. H. and DOWNES, V. A. (1968) A program for short-term forecasting, *The Statistican,* **18**, 141–147.

WALLIS, K. F. (1972) Testing for fourth order autocorrelation in quarterly regression equations, *Econometrica,* **40**, 617–636.

WALLIS, K. F. (1974) Seasonal adjustment and relations between variables, *J. Amer. Statist. Ass.,* **69**, 18–31.

WETHERILL, G. B. (1969) *Sampling Inspection and Quality Control.* London: Methuen.

WHEELWRIGHT, S. C. and MAKRIDAKIS, S. (1973) An examination of the use of adaptive filtering in forecasting, *Op. Res. Quart.,* **24**, 55–64.

WHITTLE, P. (1953) Estimation and information in stationary time series, *Arkiv. für Mathematik,* **2**, 423–434.

WHITTLE, P. (1963) *Prediction and Regulation.* London: E.U.P.

WHITTLE, P. (1969) A view of stochastic control theory, *J. R. Statist. Soc.* A, **132**, 320–334.

WIENER, N. (1949) *Extrapolation, Interpolation, and Smoothing of Stationary Time-Series.* Cambridge, Mass.: M.I.T. Press.

REFERENCES

WILSON, G. T. (1971) Recent developments in statistical process control, *Bull. of the Int. Statist. Inst.*, **44**, Book 1, 509–534.

WINTERS, P. R. (1960) Forecasting sales by exponentially weighted moving averages, *Man. Sci.*, **6**, 324–342.

WISHART, D. M. G. (1969) A survey of control theory, *J. R. Statist. Soc*, A, **132**, 293–319.

WOLD, H. O. (1965) *Bibliography on Time-Series and Stochastic Processes.* Edinburgh: Oliver and Boyd.

WOLD, S. (1974) Spline functions in data analysis, *Technometrics,* **16**, 1–11.

YAGLOM, A. M. (1962) *An Introduction to the Theory of Stationary Random Functions.* Englewood Cliffs, N.J.: Prentice-Hall.

YOUNG, P. (1974) Recursive approaches to time series analysis, *Bull. I.M.A.*, **10**, 209–224.

Answers to Exercises

Chapter 2

1. (b) There are various ways of assessing the trend and seasonal effects. The simplest method is to calculate the four yearly averages in 1967, 1968, 1969 and 1970, and also the average sales in each of periods I, II, ... , XIII (i.e. row and column averages). The yearly averages estimate trend, and the differences between the period averages and the overall average estimate the seasonal effects. With such a small downward trend, this rather crude procedure may well be adequate. It has the advantage of being easy to understand and to compute. A more sophisticated approach is to calculate a 13-month simple moving average, moving along one period at a time. This will give trend values for each period from period 7 to period 46. Extrapolate (by eye) the trend values to the end of the series. Calculate the difference between each observation and the corresponding trend value. Hence estimate the average seasonal effects in periods I, ... , XIII.

2. (d) $r_1 = -0.549$. Equation (2.4) is ideal if computation is performed on a computer. If a desk machine is used, note that

$$\sum_{t=1}^{N-1} (x_t - \bar{x})(x_{t+1} - \bar{x}) = \sum_{t=1}^{N-1} x_t x_{t+1}$$

$$- \bar{x} \left(\sum_{t=1}^{N-1} x_t + \sum_{t=2}^{N} x_t \right) + (N-1)\bar{x}^2 .$$

   For this particular set of data, $\bar{x}$ happens to be one, and it is easiest to subtract one from each observation.

3. $\bar{x} \to 0$ as $N \to \infty$. $r_k \to \dfrac{\sum \cos \omega t \cos(t + k)\omega}{\sum \cos^2 \omega t}$

Use the trigonometric result:

$$2 \cos A \cos B = \cos(A + B) + \cos(A - B).$$

4. $\pm 2/\sqrt{N} = \pm 0.1$. Thus r_7 is just 'significant', but unless there is some physical explanation for an effect at lag 7, there is no real evidence of non-randomness as one expects 1 in 20 values of r_k to be 'significant' when data are random.

Chapter 3

1.
$$\rho(k) = \begin{cases} 1 & k = 0 \\ 0.56/1.53 & k = \pm 1 \\ -0.2/1.53 & k = \pm 2 \\ 0 & \text{otherwise} \end{cases}$$

3. $\text{Var}(X_t)$ is not finite. $Y_t = Z_t + (C - 1)Z_{t-1}$.
$$\rho_Y(k) = \begin{cases} 1 & k = 0 \\ (C - 1)/[1 + (C - 1)^2] & k = \pm 1 \\ 0 & \text{otherwise} \end{cases}$$

4. $\rho(k) = 0.7^{|k|} \qquad k = 0, \pm 1, \pm 2, \ldots$

6. We must have $\left\| \dfrac{\lambda_1 \pm \sqrt{(\lambda_1^2 + 4\lambda_2)}}{2} \right\| < 1.$

If $\lambda_1^2 + 4\lambda_2 > 0$, can easily show $\lambda_1 + \lambda_2 < 1$, and $\lambda_1 - \lambda_2 > -1$. If $\lambda_1^2 + 4\lambda_2 < 0$, roots are complex and we find $\lambda_2 > -1$.

8. $\gamma_Y(k) = 2\gamma_X(k) - \gamma_X(k + 1) - \gamma_X(k - 1)$.
$$\gamma_Y(k) = \begin{cases} 2 - 2\lambda & k = 0 \\ -\lambda^{|k-1|}(1 - \lambda)^2 & k = \pm 1, \pm 2, \ldots \end{cases}$$

9. $X_t = Z_t + 0.3\, Z_{t-1} + 0.3^2\, Z_{t-2} + \ldots$.

10. If B^* denotes the backward shift operator, then $A(B^*)\, X_t = Z_t$ and $X_t = B(B^*)\, Z_t$, so that $Z_t = A(B^*)B(B^*)\, Z_t$. Thus $A(B^*)B(B^*)$ is the unit operator. Hence $A(s)B(s) = 1$. Show coefficient of s^k in $\sigma_Z^2 B(s)B(1/s)$ is equal to $\gamma_X(k)$. Use $A(s)B(s) = 1$ to show equivalence with $\sigma_Z^2/A(s)A(1/s)$.

11. First evaluate $\Upsilon(k)$. Find

$$\gamma(0) = \sigma_Z^2(1 + \beta^2 + 2\alpha\beta)/(1 - \alpha^2)$$
$$\gamma(1) = \sigma_Z^2(\alpha + \alpha\beta^2 + \alpha^2\beta + \beta)/(1 - \alpha^2)$$
$$\Upsilon(k) = \alpha\gamma(k - 1) \qquad k = 2, 3, \ldots$$

Note that $E(X_t Z_t) = \sigma_Z^2$, and $E(X_t Z_{t-1}) = (\alpha + \beta)\sigma_Z^2$.

12. $E(X(t)) = \alpha/(\alpha + \beta)$ for all t. $\mathrm{Cov}(X(t), X(t + \tau))$
$$= E[X(t)X(t + \tau)] - \alpha^2/(\alpha + \beta)^2.$$
$$E[X(t)X(t + \tau)] = \mathrm{Prob}[\text{both } X(t) \text{ and } X(t + \tau) \text{ are } 1]$$
$$= \mathrm{Prob}(X(t) = 1)\mathrm{Prob}(X(t + \tau) = 1 \mid X(t) = 1).$$

13. For convenience assume $E(X(t)) = 0$. Then

$$g(T) = \frac{1}{T^2} \int_0^T \int_0^T E[X(t)X(t')] \, dt \, dt'.$$

Make appropriate change of variable.

Note that $\dfrac{d}{dT}\left[\int_0^T g(u)du\right] = g(T)$.

14. $\mathrm{Cov}[Ye^{i\omega t}, \bar{Y}e^{-i\omega(t+\tau)}] = e^{-i\omega\tau}\mathrm{Cov}[Y, \bar{Y}]$ does not depend on t.

Chapter 4

2. $\Sigma(x_t - \bar{x})(x_{t-1} - \bar{x}) = \hat{\alpha}_1 \Sigma(x_{t-1} - \bar{x})^2 + \ldots +$
$$\hat{\alpha}_m \Sigma(x_{t-m} - \bar{x})(x_{t-1} - \bar{x})$$
. . . .
$\Sigma(x_t - \bar{x})(x_{t-m} - \bar{x}) = \hat{\alpha}_1 \Sigma(x_{t-1} - \bar{x})(x_{t-m} - \bar{x}) + \ldots +$
$$\hat{\alpha}_m \Sigma(x_{t-m} - \bar{x})^2.$$

All summations are for $t = (m + 1)$ to N. These equations are the same as the Yule–Walker equations except that the constant divisor $\Sigma(x_t - \bar{x})^2$ is omitted and that r_k is calculated from $(N - m)$ instead of $(N - k)$ cross-product terms.

3. Hint: see Section 3.4.4. $\rho(2) = \alpha_1\rho(1) + \alpha_2$; $\rho(1) = \alpha_1 + \alpha_2\rho(-1) = \alpha_1 + \alpha_2\rho(1)$. Solve for α_2.

4. Values outside $\pm2/\sqrt{100} = \pm0.2$ are 'significant'. i.e. r_1 and r_2. A second-order MA process has an ac.f. of this type.

ANSWERS TO EXERCISES

Chapter 5

2. Denote model by

$$(1 - \alpha B^{12})w_t = (1 + \beta B)a_t \text{ or}$$
$$x_t = a_t + \beta a_{t-1} + (1 + \alpha)x_{t-12} - \alpha x_{t-24}.$$
Then $\hat{x}(N,1) = (1 + \hat{\alpha})x_{N-11} - \hat{\alpha}x_{N-23} + \hat{\beta}\hat{a}_N$
and $\hat{x}(N,k) = (1 + \hat{\alpha})x_{N+k-12} - \hat{\alpha}x_{N+k-24}$ $(k = 2, 3, \ldots, 12.)$

Chapter 6

1. (a) $f(\omega) = \sigma_Z^2/\pi(1 - 2\lambda \cos \omega + \lambda^2)$
   (b) $f^*(\omega) = (1 - \lambda^2)/\pi(1 - 2\lambda \cos \omega + \lambda^2)$

2. (a) $f(\omega) = \sigma_Z^2 [3 + 2(2 \cos \omega + \cos 2\omega)]/\pi$
   (b) $f(\omega) = \sigma_Z^2 [1.34 + 2(0.35 \cos \omega - 0.3 \cos 2\omega)]/\pi$.

3. The non-zero mean makes no difference to the acv.f., ac.f. or spectrum.

$$\gamma(k) = \begin{cases} 1.89 \, \sigma_Z^2 & k = 0 \\ 1.2 \, \sigma_Z^2 & k = \pm 1 \\ 0.5 \, \sigma_Z^2 & k = \pm 2 \\ 0 & \text{otherwise} \end{cases}$$

$$\rho(k) = \begin{cases} 1 & k = 0 \\ 1.2/1.89 & k = \pm 1 \\ 0.5/1.89 & k = \pm 2 \\ 0 & \text{otherwise} \end{cases}$$

4. $$\rho(k) = \int_0^\pi f^*(\omega) \cos \omega k \, d\omega$$

$$= -\frac{2 \cos \omega k}{\pi^2 k^2} \Big]_0^\pi$$

5. Clearly $E(X(t)) = 0$
$E[X(t)X(t + u)] = \text{Prob}[X(t) \text{ and } X(t + u) \text{ have same sign}]$
$\qquad - \text{Prob}[X(t) \text{ and } X(t + u) \text{ have opposite sign}]$

Hint: Prob(even number of changes in time u)

$$= e^{-\lambda u}\left[1 + \frac{(\lambda u)^2}{2!} + \frac{(\lambda u)^4}{4!} + \dots\right]$$

$$= e^{-\lambda u}(e^{\lambda u} + e^{-\lambda u})/2.$$

Chapter 7

6. $\quad 2N\Big/\sum\limits_{k=-M}^{+M} \frac{1}{4}\left(1 + \cos\frac{\pi k}{M}\right)^2 = 8N/(3M+1) \to 8N/3M$ as $M \to \infty$.

Chapter 8

1. $\quad f_{xy}(\omega) = \dfrac{\sigma_Z^2}{\pi}\left[\beta_{11}\beta_{21} + \beta_{12}\beta_{22} + \beta_{21}e^{-i\omega} + \beta_{12}e^{+i\omega}\right].$

2. Use the fact that $\text{Var}[\lambda_1 X(t) + \lambda_2 Y(t+\tau)] \geqslant 0$ for any constants λ_1, λ_2.

$$\gamma_{xy}(k) = \begin{cases} 0.84\ \sigma_Z^2 & k = 0 \\ -0.4\ \sigma_Z^2 & k = 1 \\ +0.4\ \sigma_Z^2 & k = -1 \\ 0 & \text{otherwise} \end{cases}$$

$c(\omega) = 0.84\ \sigma_Z^2 /\pi$

$q(\omega) = -0.8\ \sigma_Z^2 \sin\omega/\pi$

$\alpha_{xy}(\omega) = \dfrac{\sigma_Z^2}{\pi}\ \sqrt{(0.84^2 + 0.8^2 \sin^2\omega)}$

$\tan\phi_{xy}(\omega) = 0.8\sin\omega/0.84$

$C(\omega) = 1$

Chapter 9

1. Selected answers only.

(a) $\quad h_k = \begin{cases} \frac{1}{2} & k = \pm 1 \\ 1 & k = 0 \\ 0 & \text{otherwise} \end{cases}$

ANSWERS TO EXERCISES

$$S_t = \begin{cases} 0 & t < -1 \\ \tfrac{1}{2} & t = -1 \\ 1\tfrac{1}{2} & t = 0 \\ 2 & t \geqslant 1 \end{cases}$$

$H(\omega) = \tfrac{1}{2}e^{-i\omega} + 1 + \tfrac{1}{2}e^{i\omega} = 1 + \cos \omega$

$G(\omega) = H(\omega)$ since $H(\omega)$ is real. $\phi(\omega) = 0$.

(b) $H(\omega) = \dfrac{1}{5} + \dfrac{2}{5}\cos \omega + \dfrac{2}{5}\cos 2\omega$

(c)
$$h_k = \begin{cases} +1 & k = 0 \\ -1 & k = +1 \\ 0 & \text{otherwise} \end{cases}$$

(a) and (b) are low-pass filters. (c) and (d) are high-pass filters. The combined filter has

$$H(\omega) = (1 + \cos \omega)\left(\frac{1}{5} + \frac{2}{5}\cos \omega + \frac{2}{5}\cos 2\omega \right).$$

2. (a) $H(\omega) = ge^{-i\omega\tau}$ $(\omega > 0)$

   (b) $H(\omega) = g(1 - i\omega T)/(1 + \omega^2 T^2)$ $(\omega > 0)$.

4. $$\gamma(k) = \begin{cases} \alpha^{|k/2|}\, \sigma_Z^2/(1 - \alpha^2) & k \text{ even} \\ 0 & k \text{ odd}. \end{cases}$$

$f_X(\omega) = f_Z(\omega)/(1 - 2\alpha \cos 2\omega + \alpha^2).$ $(0 < \omega < \pi).$

257

Index

AUTHOR INDEX

Ackroyd, M. H., 230, 242
Akaike, H., 74, 103, 223-225, 242, 248
Anderson, T. W., 1, 10, 95, 129, 135, 160, 164, 242
Armstrong, J. S., 100, 242
Astrom, K. J., 94, 108, 222, 223, 228, 229, 242

Barksdale, H. C., 180, 242
Barnard, G. A., 242
Bartlett, M. S., 34, 37, 67, 107, 114, 127, 137, 151, 242, 243
Bates, J. M., 83, 243
Bather, J. A., 229, 230, 243
Battin, R. H., 108, 228, 247
Bell, D. J., 228, 243
Bendat, J. S., 11, 146, 243
Besag, J., 232, 243
Bhattacharyya, M. N., 97, 101, 243
Bingham, C., 146, 149, 243
Blackman, R. B., 10, 234, 243
Blake, I. F., 232, 243
Bloomfield, P., 94, 243
Bogert, B. P., 230, 243
Bohlin, T., 223, 229, 242
Box, G. E. P., 9, 11, 16, 21, 22, 43, 44, 49, 51, 65, 71-80, 85, 86, 89, 91, 96, 97, 174, 215, 219-225, 229, 243
Brass, W., 4, 9, 244
Brillinger, D. R., 10, 139, 231, 244
Brown, R. G., 9, 84, 85, 93, 94, 96, 244
Button, K. J., 16, 244

Caines, P. E., 229, 244
Chambers, J. C., 82, 244
Chatfield, C., 4, 43, 79, 89, 94, 163, 182, 244
Christ, C. F., 97, 244
Clarke, D. W., 223, 244
Cliff, A. D., 232, 244
Cochrane, D., 96
Coen, P. J., 174, 244
Cooley, J. W., 146, 244, 245
Cox, D. R., 5, 16, 34, 40, 49, 53, 54, 87, 107, 108, 113, 234, 243, 245
Coutie, G. A., 84, 87, 100, 245
Craddock, J. M., 163, 245
Cramer, H., 232, 245

Daniell, P. J., 143
Davis, E. J., 83, 245
Dhrymes, P. J., 245
Douce, J. L., 228, 236, 245
Downes, V. A., 250
Durbin, J., 22, 71, 76, 78, 79, 245

Enochson, L., 11, 146, 248
Eykhoff, P., 222, 223, 242

Fenton, R., 149
Fishman, G. S., 245
Freund, J. E., 240
Friedland, B., 186, 198, 238, 249
Fuller, A. T., 228, 245

258

INDEX

Gearing, H. W. G., 100, 245
Godfrey, M. D., 139, 147, 231, 232, 243, 246
Gomme, E. J., 244
Goodman, N. R., 218, 246
Granger, C. W. J., 1, 11, 13, 83, 93, 96, 99, 131, 135, 162, 164, 170, 176, 178, 184, 223, 225, 243, 246, 248
Gregg, J. V., 16, 84, 246
Grenander, U., 10, 107, 111, 246
Griffiths, J. W. R., 246
Grohman, M. C., 100, 242
Groves, G. W., 250
Gudmundsson, G., 218, 246
Guffey, H. J. Jr., 180, 242

Hamming, R. W., 143
Hammond, J. K., 230, 246
Hamon, B. V., 180, 222, 246
Hann, Julius Von, 142
Hannan, E. J., 10, 74, 135, 141, 146, 150, 162, 180, 222, 231, 246
Harris, B., 11, 246
Harrison, P. J., 16, 84, 85, 94, 246, 247
Hasselmann, K. F., 250
Hatanaka, M., 11, 131, 135, 170, 176, 178, 223, 225, 246
Haubrich, R. A., 180, 247
Hause, J. C., 200, 218, 247
Healy, M. J. R., 243
Hepple, L. W., 232, 250
Holt, C. C., 85
Hossell, C. V., 246
Hsu, H. P., 233, 235, 239, 247
Hughes, A. O., 1, 13, 162, 164, 184, 246

Jacobs, O. L. R., 228, 247
Jenkins, G. M., 9-11, 21, 22, 39, 43, 44, 49, 51, 53, 61, 65-80, 85, 86, 89, 91, 97, 111, 118, 123, 124, 129, 136, 141, 148, 149, 151, 154, 163, 169, 173, 182, 184, 198, 202, 215, 218-225, 229, 234, 242, 243, 247
Jones, R. H., 145, 151, 152, 231, 247

Katz, S., 246
Keay, F., 82, 247
Kendall, M. G., 11, 12, 17, 18, 31, 62, 71, 77, 84, 96, 119, 137, 171, 176, 244, 247
Khintchine, A. Y., 114
Kolmogorov, A. N., 107, 228
Kramer, B. H., 246
Kuo, M. T., 246

Laning, J. H., 108, 228, 247
Leadbetter, M. R., 232, 245
Lee, Y. W., 175, 247
Leuthold, R. M., 97, 247
Lewis, P. A. W., 5, 244, 245
Lindsey, W. C., 232, 243
Loynes, R. M., 230, 247

MacCormick, A. J. A., 247
MacGregor, J. F., 223-225, 229, 243, 247
MacKay, D. B., 164, 247
MacKenzie, G. S., 180, 247
Makridakis, S., 94, 250
Mann, H. B., 67
Markov, A. A., 45
Miller, G. R., 250
Miller, H. D., 34, 40, 49, 53, 54, 107, 108, 113, 234, 245
Mooers, C. N. K., 163, 248
Mullick, S. K., 244
Munk, W. H., 163, 248, 250
Musgrave, J. C., 249
Myers, C. L., 100, 248

Nakagawa, T., 248
Naylor, T. H., 89, 99, 107, 248
Neave, H. R., 141, 150, 151, 231, 248
Nelson, C. R., 89, 248
Nerlove, M., 78, 248
Newbold, P., 93, 94, 96, 99, 174, 243, 244, 246, 248
Noll, A. M., 230, 248

Orcutt, G. H., 96
Ord, J. K., 232, 244
Otnes, R. K., 11, 146, 248
Otomo, T., 225, 248

259

INDEX

Papoulis, A., 34, 248
Pagano, M., 68, 248
Parzen, E., 34, 178, 248, 249
Pearce, S. F., 16, 84, 247
Pepper, M. P. G., 163, 182, 244
Pierce, D. A., 78, 243, 249
Piersol, A. G., 11, 146, 243
Plackett, R. L., 229
Pocock, S. J., 129, 131, 249
Powers, W. H., 250
Priestley, M. B., 223, 224, 230, 249
Prothero, D. L., 4, 43, 79, 89, 106, 244

Quenouille, M. H., 10, 249

Rappoport, J. Q. G. H., 250
Reid, D. J., 94, 99, 249
Richardson, J. T., 246
Robinson, E. A., 11, 249
Roden, G. I., 163, 249
Rosenblatt, M., 10, 11, 107, 111, 246, 249

Sande, G., 146
Schmitz, A., 247
Schuster, A., 127
Schwarz, R. J., 186, 198, 238, 249
Seaks, T. G., 248
Shapiro, H. S., 231, 249
Shiskin, J., 22, 249
Silverman, R. A., 231, 249
Singleton, R. C., 147, 249
Smith, D. D., 244
Smith, R. L., 163, 248
Sneddon, I. H., 233, 236, 250
Snodgrass, F. E., 163, 248, 250
Solodovnikov, V. V., 228, 236, 250
Stern, G. J. A., 101
Stevens, C. F., 94, 247
Stocklin, P. L., 246
Stuart, A., 11, 12, 18, 31, 62, 119, 137, 171, 176, 247

Tee, L. H., 225, 250
Tetley, H., 17, 250
Thompson, H. E., 89, 250
Tiao, G. C., 89, 250
Tick, L. J., 184, 250
Tomasek, O., 103, 250
Tong, H., 230, 249
Tucker, M. J., 248
Tukey, J. W., 10, 127, 146, 147, 230, 234, 243, 250

Unwin, D. J., 232, 250

Wagle, B., 100, 250
Wald, A., 67
Walker, Sir Gilbert, 48
Wallis, K. F., 77, 78, 248, 250
Watson, G. S., 76, 245
Watts, D. G., 10, 39, 53, 61, 66, 68, 69, 73, 111, 118, 123, 124, 129, 136, 141, 148, 149, 151, 154, 163, 169, 173, 182, 184, 198, 202, 218, 222, 234, 247
Welch, P. D., 244, 245
Wetherill, G. B., 5, 250
Wheelwright, S. C., 94, 250
Whittle, P., 11, 70, 107, 229, 250
Wichern, D. W., 248
Wiener, N., 11, 107, 108, 113, 114, 228, 250
Wilson, G. T., 223, 251
Winsten, C. B., 242
Winters, P. R., 87, 251
Wishart, D. M. G., 229, 251
Wold, H. O., 11, 54, 90, 251
Wold, S., 18, 231, 251
Wu, S. M., 225, 250

Yaglom, A. M., 34, 53, 59, 65, 107, 108, 235, 251
Young, A. M., 249
Young, P., 222, 229, 230, 251
Yule, G. U., 44, 48, 90

260

SUBJECT INDEX

Abbreviations, xv
Adaptive filtering, 94
Additive seasonal effect, 15, 22, 88
Aggregated series, 6
Aliasing, 156, 157
Alignment, 182, 184
Alternating series, 27
Amplitude, 111, 134
Angular frequency, 111
Autocorrelation coefficient, 23-30
Autocorrelation function, 10
    of an AR process, 46, 48
    of an MA process, 42
    of a purely random process, 39
    definition of, 36
    estimation of, 60-65, 155
    properties, 37-39
Autocovariance coefficient, 24, 25, 36,
    60
    circular, 149
Autocovariance function, 35-37, 117
    estimation of, 60-62, 148
Automatic forecasting procedure, 89,
    99
Autoregressive process, 44-50
    estimation for, 65-70
    spectrum of, 207

Back-forecasting, 91
Band-limited, 156
Bandwidth, 154
Bartlett window, 141
Beveridge series, 1, 13, 162
Binary process, 5
Bispectrum, 231
Bivariate process, 169
Box–Cox transformation, 16
Box–Jenkins approach, 11, 219
Box–Jenkins control, 9, 229
Box–Jenkins forecasting, 89-93,
    97, 99-101, 103, 106
Box–Jenkins seasonal model, 74-76,
    91
Brown's method, 94, 99

Cepstrum, 231
Closed-loop system, 223
Coherency, 177, 218
Combination of forecasts, 83
Continuous process, 52, 119
Continuous time series, 6
Continuous trace, 2, 155, 158
Control theory, 9, 11, 228-230
Convolution, 20
Correlation coefficient, 23, 240
Correlogram, 25-30, 63-65
Co-spectrum, 176
Covariance, 35, 240
Cross-amplitude spectrum, 177
Cross-correlation function, 169-172
    estimation of, 172-174
Cross-covariance function, 169-172
    estimation of, 172-174
Cross-periodogram, 183
Cross-spectrum, 174-180
    estimation of, 180-184, 222
Curve-fitting, 16

Delay, 172, 180, 189, 190, 202, 220
Delayed exponential response, 190
Delphi method, 82
Delta function, 238
Deterministic process, 54
Deterministic series, 6
Diagnostic checking, 90, 91
Differencing, 21, 22, 90, 212, 223
Dirac delta function, 124, 190, 238,
    239
Discounted least squares, 94
Discrete time series, 6
Dispersive case, 180
Distributed lag model, 219
Double exponential smoothing, 94
Durbin–Watson statistic, 77

Echo, 231
Econometric model, 97, 99
Economic time series, 1, 162, 163
Endogenous variable, 78, 97

INDEX

Ensemble, 34
Equilibrium distribution, 36
Ergodic theorems, 65
Evolutionary spectrum, 230
Exogenous variable, 97
Exponential response, 190, 201
Exponential smoothing, 18, 85-89
Extrapolation, 107

Fast Fourier transform, 145-149
Feedback, 223-226
Filter, 17, 211, 212
Filtering, 108, 228
Forecasting, 9, 82-108
Fourier analysis, 127, 133, 145
Fourier transform, 118, 198, 233-237
Frequency, 111
Frequency domain, 10, 110
Frequency response function, 193-198
    estimation of, 215-219

Gain, 177, 189, 195, 198-200
General exponential smoothing, 94
General linear process, 52
Gompertz curve, 16, 84

Hamming, 143
Hanning, 142
Harmonic, 139, 164, 231
Harmonic analysis, 133
High frequency spectrum, 121
High-pass filter, 19, 200, 212
Hill-climbing, 72, 73
Holt—Winters procedure, 87-89, 99,
    102, 103, 106

Impulse response function, 187-192
    estimation of, 219-223
Integrated model, 51, 74
Invertibility, 43

Jackknife estimation, 61, 62

Kalman filtering, 94, 229, 230
Kernel, 151
Kolmogorov—Smirnov test, 79

Lag, 36
Lag window, 139
Langevin equation, 53
Laplace transform, 196, 198, 235-237
Lead time, 82, 83
Leading indicator, 83, 106
Line spectrum, 134
Linear filter, 17
Linear system, 8, 10, 186-212
    identification, 213-226
Logarithmic transformation, 15
Logistic curve, 16
Low frequency spectrum, 121
Low-pass filter, 19, 198, 212

Marketing time series, 4, 163
Markov process, 45, 126
Mean, 35
Missing observations, 231
Mixed model, 50
    estimation for, 73, 74
Model-building, 79
Moving average, 17
Moving average process, 41-44
    estimation for, 70-73
    spectrum of, 206
Multiple regression, 8, 95-97
Multiplicative seasonal effect, 15, 22,
    88
Multivariate forecast, 83, 95-101
Multivariate normal distribution, 37

Noise, 108, 228
Non-linear estimation, 72
Non-stationary series, 27, 230
Normal process, 37, 39
Normalized spectral density function,
    118
Nyquist frequency, 113, 131, 132

Open-loop system, 223
Ornstein—Uhlenbeck process, 54
Outlier, 7, 30

Parseval's theorem, 134
Parzen window, 140, 150, 153

262

INDEX

Partial autocorrelation function, 64, 68, 69
Periodogram, 135-139
Phase, 111, 134, 195, 198-200
Phase spectrum, 177
Physically realizable, 187
Point process, 5
Polynomial curve, 16, 18, 84
Polyspectra, 231
Power, 115
Power spectrum, 116
Prediction, 8, 108
Prediction model, 83
Prediction theory, 107, 108
Prewhitening, 160
Process control, 4
Projection, 83, 84
Purely indeterministic, 54, 115
Purely random process, 39-41

Quadrature spectrum, 176

Random walk, 40, 52
Realization, 34
Residuals, 16, 76, 91, 96
Resolution, 141, 144
Resonant, 187

Sample spectrum, 136
Sampled series, 6
Seasonal variation, 7, 12, 14, 17, 21, 22, 28, 30
Second-order stationary, 37
Serial correlation coefficient, 24
Signal, 108, 228
Slutsky effect, 19
Smoothing constant, 86
Spatial series, 232
Spectral analysis, 10, 127-168
Spectral density function, 10, 110, 116-120
   of an AR process, 121
   of an MA process, 120
   estimation of, 127-168
Spectral distribution function, 111-117

Spectral representation, 113, 114
Spectral window, 151
Spectrogram, 135
Spectrum, 116, 127
Spencer's 15 point moving average, 17, 20, 212
Spline function, 18, 231
Stable system, 188
Stationary distribution, 36
Stationary process, 35
Stationary time series, 14
Steady state, 196
Step change, 94, 139
Step response function, 192
Stochastic process, 6, 33
Strictly stationary, 35
Subjective forecasts, 82, 100

Tapering, 147
Time domain, 10, 60
Time invariance, 187
Time Plot, 14
Transfer function, 193, 198
Transfer function model, 97, 220
Transformations, 14
Transient, 196
Trend, 7, 12-18, 21, 27
Trend curve, 84
Truncation point, 140
Tukey window, 140, 150, 153
Turning point, 8

Unequal intervals, 231
Univariate forecast, 82, 84-94, 99-101

Variance, 35

Wavelength, 111
Weakly stationary, 37
White noise, 39
Wiener—Hopf equation, 216
Wiener—Khintchine theorem, 114, 118
Wold decomposition, 54, 115

Yule—Walker equations, 48, 68

Z-transform, 198, 236